Textbook of

Nanoscience

and

Nanotechnology

Textbook of Nanoscience and Nanotechnology

B S Murty
Professor,
Indian Institute of Technology Madras, Chennai, India

P Shankar
Principal,
Saveetha Engineering College, Chennai, India

Baldev Raj
President–Research
PSG Institutions, Coimbatore, India

B B Rath
Director,
Materials Science and Component Technology,
Naval Research Laboratory, Washington DC, USA

James Murday
Scientist,
Naval Research Laboratory, Washington DC, USA

Editor-in-Chief: Baldev Raj

IIM Metallurgy Materials Engineering

Series in Metallurgy and Materials Science

Springer

Universities Press

B S Murty
Professor
Indian Institute of Technology Madras
Chennai, India

P Shankar
Principal
Saveetha Engineering College
Chennai, India

Baldev Raj
President–Research
PSG Institutions, Coimbatore, India

B B Rath
Director
Materials Science and
Component Technology
Naval Research Laboratory
Washington DC, USA

James Murday
Scientist
Naval Research Laboratory
Washington DC, USA

e-ISBN 978-3-642-28030-6
DOI 10.1007/978-3-642-28030-6

Bangalore / Bhopal / Bhubaneshwar / Chandigarh / Chennai
Ernakulam / Guwahati / Hyderabad / Jaipur / Kolkata
Lucknow / Mumbai / New Delhi / Noida / Patna

Typeset in MinionPro 11/13 by
OSDATA, Hyderabad 500 029

Foreword

The emergence of nanoscience and nanotechnology has opened a myriad possibilities to revolutionize a wide range of fields ranging from cosmetics to space technologies; there are also important contributions to diagnostics and therapeutic practices, especially for treating cancer. Nanoscience is also making a difference in many areas of engineering and technology, such as energy environmental pollution control, textile, automobile and electronics. Nanoscience and nanotechnology combine the essence of knowledge in many fields of science and are truly interdisciplinary. There is little doubt that the next generation of citizens will be the beneficiaries of the applications of nanoscience and nanotechnology in their daily life. It is, therefore, vital to have adequate awareness of this important field, irrespective of the specialization or profession that one may choose.

This book presents a fine insight into many aspects of nanoscience and their applications in engineering. The book has been written from the perspective of materials scientists, but the presentation is lucid and simple and easily understandable by beginners and undergraduate students. The valuable experience of the authors as teachers and scientists is borne out by the balanced perspective provided by the text. While emphasizing the applications of nanotechnology that are yet to be explored, the authors have also indicated possible issues related to environmental and ecological concerns.

I trust that students, research scholars and teachers will find this book a useful introduction to nanoscience and nanotechnology.

C N R Rao
National Research Professor and Honorary President
Jawaharlal Nehru Centre for Advanced Scientific Research
Bengaluru, India

Preface

The impact of anything associated with 'nano' can be compared to Vamana, a well-known figure in Indian Mythology; though diminutive in size, he encompassed the entire universe with one tiny step! Similarly, nanoscience's omnipresence can be felt in almost all fields of engineering and technology. Since the Industrial Revolution, Manhattan Project (utilization of nuclear energy), deployment of Sputnik (led to space exploration) and ARPANET (resulted in information explosion), it is the only science that has brought together people of different fields, making it truly inter-disciplinary. 'Nano' as a term is no longer restricted to research laboratories and institutes; it is seen everywhere, in applications ranging from consumer goods such as toothpaste, wall paints, water purifiers and refrigerators to sophisticated applications such as catalysis, sensors and electronics.

Though many books on nanoscience and nanotechnology are available, very few have been written with the beginner in mind. This book attempts to bridge this gap. The presentation is kept simple so that beginners and undergraduate students will be able to understand and obtain a good overview of the subject. To kindle interest in this field in the minds of the readers, brief biographical sketches of people who contributed significantly to this field have been given. Students and research scholars in this field will find this book immensely useful. The science and technology of regulating the controlled application of this revolutionary technology for restraining the negative impacts is perhaps going to be the biggest challenge of the future.

The first chapter introduces the reader to the nano-world and brings out the exciting developments in this field. The second chapter identifies the unique properties of materials at the nanoscale. The synthesis routes, both bottom-up and top-down, have been discussed briefly in the third chapter. The applications of nanomaterials, ranging from medical to electronic, have been highlighted in the fourth chapter. The fifth chapter gives a brief description of the various tools used to characterize materials at the nanoscale. The most widely studied nanostructures such as quantum dots, carbon nanotubes, fullerenes and graphenes, nanowires, etc., are presented in the sixth chapter. The last chapter brings out the health hazards related to nanomaterials.

In this book, we have made special efforts to showcase the contributions of Indian scientists in the field of nanoscience and nanotechnology. We are grateful to Dr N Ravisankar, IISc Bangalore, Dr Bhaskar Majumdar, DMRL, Hyderabad, Prof. V Ramgopal Rao, IIT Bombay, Prof. M S Ramachandra Rao, Prof. Uday Chakkingal, Prof. T S Sampath Kumar and Dr R Sarathi, IIT Madras, who have contributed their research work to enrich this book. Prof. Murty would specially like to thank his research scholars, both past and present, Joydip, Monikanchan, Kallol, Venugopal, Parashar, Barua, Sashank, Jatin, Varalakshmi, Shanmugasundaram, Udhayabanu, Srinivasulu, Ajeet, Ramakrishna, Murugan, Praveen Kumar, Karthikeyan, Anisha, Praveen, Susila and Prakash, all of who have significantly contributed to this field and enriched his knowledge. Some parts of their work have been referred to in this book, which is gratefully acknowledged. Discussions and associations with experts in this field, such as Profs S Ranganathan, K Chattopadhyay, T Pradeep, S Ramaprabhu, S K Das, I Manna, S K Pabi and C Suryanarayana, and Drs A K Tyagi, G K Dey and B L V Prasad, are gratefully acknowledged.

The authors acknowledge the enthusiastic support and meticulous editorial coordination of Universities Press at various stages of the preparation of the book. A special word of appreciation to Ms Javanthi Singaram and Ms Madhavi Sethupathi for their untiring editorial support. We welcome suggestions from readers towards improvements that can be incorporated in future editions of this book.

B S Murty, P Shankar, Baldev Raj, B B Rath and **James Murday**

About the Series

The study of metallurgy and materials science is vital for developing advanced materials for diverse applications. In the last decade, the progress in this field has been rapid and extensive, giving us a new array of materials, with a wide range of applications, and a variety of possibilities for processing and characterizing the materials. In order to make this growing volume of knowledge available, an initiative to publish a series of books in Metallurgy and Materials Science was taken during the Diamond Jubilee year of the Indian Institute of Metals (IIM) in the year 2006. As part of the series we have already brought out five books, and all of them have been copublished by CRC Press, USA, for distribution overseas. This is the sixth book in the series and the second textbook to be published.

The IIM is a premier professional body representing an eminent and dynamic group of metallurgists and materials scientists from R&D institutions, academia and industry in India. It is a registered professional institute with the primary objective of promoting and advancing the study and practice of the science and technology of metals, alloys and novel materials.The institute is actively engaged in promoting academia–research and institute–industry interactions.

Universities Press, an associate of Orient Blackswan, with its long tradition of publication of quality books in engineering and sciences, has come forward to undertake the publication of this series, thus synergising the professional expertise of IIM with the publishing experience of Universities Press towards effective knowledge dissemination. This book series shall include different categories of publications: textbooks to satisfy the requirements of undergraduates and beginners in the field, monographs on select topics by experts in the field, and proceedings of select international conferences organized by IIM after mandatory peer review. To increase the readership and to ensure wide dissemination, some of the books in the series will be copublished with international publishers.

The international character of the authors and editors has helped the books command a global readership. An eminent panel of international and national experts acts as the advisory body in overseeing the selection of topics, important areas to be covered, and the selection of contributing authors. These publications are expected to serve as a source of knowledge to a wide spectrum of students, engineers, researchers and industrialists in the field of metallurgy and materials science. I look forward to receiving your valuable response to the present book as well as the other books in the series.

Baldev Raj
Editor-in-Chief

About the Authors

B S Murty, Professor, Department of Metallurgical and Materials Engineering, IIT Madras, Chennai, India, has pioneered the synthesis of nanomaterials by high-energy ball milling/mechanical alloying route and has also made significant contributions to the field of bulk metallic glasses and their nanocrystallization. His other fields of interest are in-situ composites and metal foams.

P Shankar, Principal, Saveetha Engineering College, Chennai, India, is a prolific researcher and teacher with National and International credentials. As senior research scientist at the Department of Atomic Energy, he made several pioneering contributions in the field of characterization of advanced nuclear materials and in the development of novel surface engineering processes that are being used in our nuclear reactors.

Baldev Raj, President–Research, PSG Institutions, Coimbatore, India, has made pioneering contributions in the area of non-destructive characterization, evaluation of defects, microstructures and stresses in structural and strategic materials. His contributions also encompass the design and development of nuclear materials such as austentic steels, low-activation martensitic/ferritic steels and nuclear fuels, which have a wide range of applications in fission and fusion reactors, advanced clean energy systems and aerospace structures. He is a leading authority in materials, welding, corrosion and structure–property correlation studies. His work in nanoscience and technology relates to nanofluids, nanosensors and coatings.

Bhakta B Rath, Associate Director, Materials Science and Component Technology Directorate, Naval Research Laboratory, Washington DC, USA, is an educator, researcher and administrator. His research interests are in the fields of solid state transformations and structure–property relationships. He serves on the planning and advisory boards of a number of US Government agencies, academies and universities.

James S Murday, Director of Physical Sciences, University of Southern California's Research Advancement Office in Washington DC, USA, led the Naval Research Laboratory's Chemistry Division for twenty years. He served as Director of the US National Nanotechnology Coordination Office in 2001–2002 and as Executive Secretary to the US National Science and Technology Council's Subcommittee on Nanometer Science Engineering and Technology (NSET) in 2001–2006.

List of Colour Plates

Contents

Chapter 1

The Big World of Nanomaterials

Learning objectives

- Introduction to nanocrystalline materials
- Brief history of research in nanomaterials
- Exciting developments in the science and technology of nanomaterials
- Classification of nanomaterials

Materials have been of great interest to human beings since time immemorial. A few million years ago, it was found that rocks could be used to break things that were impossible to break with bare hands. Stones were the first tools and even today they are still in use in kitchens and laboratories to pound and grind, or as mortars and pestles. Around 5000–6000 years ago, it was accidentally discovered that when a rock containing copper was placed on a fire, molten copper could be collected. This discovery led to the reduction of metal ores to produce metals for the fabrication of items from ploughshares to swords. New materials with greater hardness and longer use than stone became available for making tools. Our growth and progress have paralleled the development of metals and metallurgy.

Traditionally, civilizations were named after the metals or materials used. Thus, we have the Stone Age, Bronze Age and Iron Age (Fig. 1.1). The current age is driven by the applications

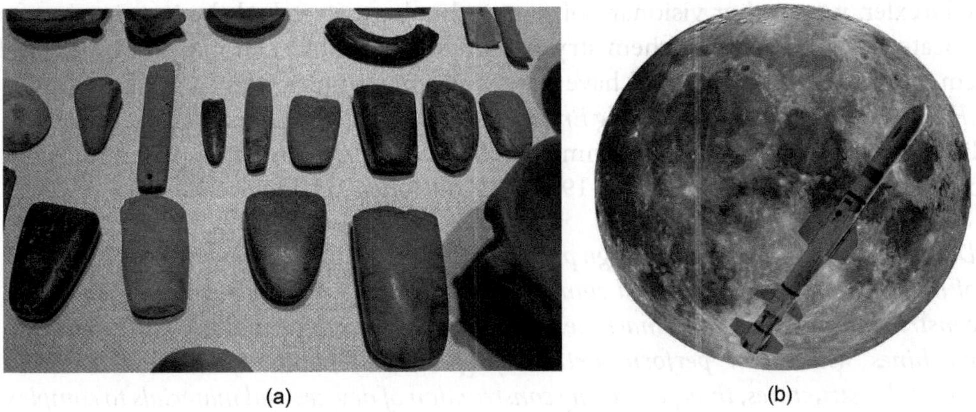

(a) (b)

Fig. 1.1 Materials evolution—from the Stone Age (a) to the Space Age (b)
(*Source:* http://rubens.anu.edu.au/raider5/greece/thessaloniki/museums/archaeological/neolithic/thessaloniki/museums/archaeological/neolithic).

of silicon and other advanced materials. The coming decades could well be dominated by *nanotechnology* ('Nano Age'), a deviation from the practice of identification of an era based on materials to one based on technology.

New technologies require new materials with superior physical, chemical and mechanical properties. Materials science and engineering have provided us materials with widely varying properties made by changing the composition or altering the microstructure using thermo-chemical-mechanical methods. Consequently, microstructural engineering and the study of structure–property correlation have become very important. The mechanism by which ultrafine microstructures affect the properties of solids could be better understood after the advent of the theory of lattice defects and dislocation theory, and the availability of advanced high-resolution microscopy techniques such as electron, atomic force and field ion microscopy. These developments have helped in understanding the correlation between the structure and properties of solids.

The unique properties of materials due to ultrafine particle sizes were recognized early in the 20th century. The classic lecture by Richard P Feynman titled "There's plenty of room at the bottom", on 29 December 1959, at the annual meeting of the American Physical Society, opened up a whole new field, known as 'nanotechnology'. He spoke about manipulating and controlling things on a small scale (For full text: http://www.zyvex.com/nanotech/feynman.html). He stated:

> *In the year 2000, when they look back at this age, they will wonder why it was not until the year 1960 that anybody began seriously to move in this direction.*

Due to his foresight, Feynman is often regarded as the first visionary of nanotechnology. Feynman's profound vision resulted in a lot of discussion and interest among the research community on engineering at a very small scale. However, it took almost three decades for the research community to translate his vision into reality for want of proper tools and techniques.

Eric Drexler, yet another visionary of nanotechnology, stretched the imagination further. He advocated that the power of chemistry be used to build molecular machines, and predicted the tremendous impact this would have on a wide spectrum of technologies. In his popular book, *Engines of Creation: The Coming Era of Nanotechnology*, he speaks about the power of an interdisciplinary approach to make commendable progress in nanoscience and nanotechnology. The following statement by Drexler in 1986 still continues to be relevant.

> *Development of the ability to design protein molecules will, by analogy between features of natural macromolecules and components of existing machines, make possible the construction of molecular machines. These machines can build second-generation machines which can perform extremely general synthesis of three-dimensional molecular structures, thus permitting construction of devices and materials to complex atomic specifications. This capability has implications for technology in general and in particular for computation and characterization, manipulation, and repair of biological materials.*

1.1 HISTORY AND SCOPE

The word '*nano*' is to a Greek prefix meaning dwarf or something very small and depicts one billionth (10^{-9}) of a unit (refer Table 1.1). Nanomaterials, therefore, refer to the class of materials with at least one of the dimensions in the nanometric range.

How small is a nanocrystal? For an immediate comparison, a nanometre represents a dimension about a few tens of thousand times thinner than human hair. Figure 1.2 gives an idea of the scale of different objects, from macroscale to nanoscale. In the case of polycrystalline materials, the grain size is typically of the order of 1–100 microns (1 micron = 10^{-6} m). Nanocrystalline materials have a grain size of the order of 1–100 nm, and are therefore 100–1000 times smaller than conventional grain dimensions. However, compared to the size of an atom (0.2–0.4 nm in diameter), nanocrystalline grains are still significantly big.

For example, a nanocrystal of size 10 nm contains over a hundred thousand atoms (assuming a spherical nanograin of 10 nm and atomic diameter of 0.2 nm), large enough to exhibit bulk properties, i.e., properties different from those of single atoms or clusters (see box on page 5 for details). As the dimensions reduce to below 50–100 nm, they can no longer be treated as infinite systems and the resultant boundary effects lead to fascinating and useful properties, which can be explored and tailored for a variety of structural and functional applications.

In principle, one cannot define an exact dimension for the grain size below which materials can be classified as 'nano'. This is because it is subjective and depends on the application or end property of interest. Most electronic and optical properties vary when the grain size is reduced typically below 10 nm. However, their mechanical, chemical and many physical properties begin to vary significantly from bulk below 50–100 nm. Hence, ***nanomaterials may be classified as those materials which have at least one of their dimensions in the nanometric range, below which there is significant variation in the property of interest compared to microcrystalline materials.***

Nanomaterials can be metals, ceramics, polymers or composites. Nanotechnology is an umbrella term for many areas of research dealing with objects that have one of their dimensions

Table 1.1 The world of small dimensions

Number	Name	Symbol
0.1	deci	d
0.01	centi	c
0.001	milli	m
0.000 001	micro	μ
0.000 000 001	nano	n
0.000 000 000 001	pico	p
0.000 000 000 000 001	femto	f
0.000 000 000 000 000 001	atto	a
0.000 000 000 000 000 000 001	zepto	z
0.000 000 000 000 000 000 000 001	yocto	y

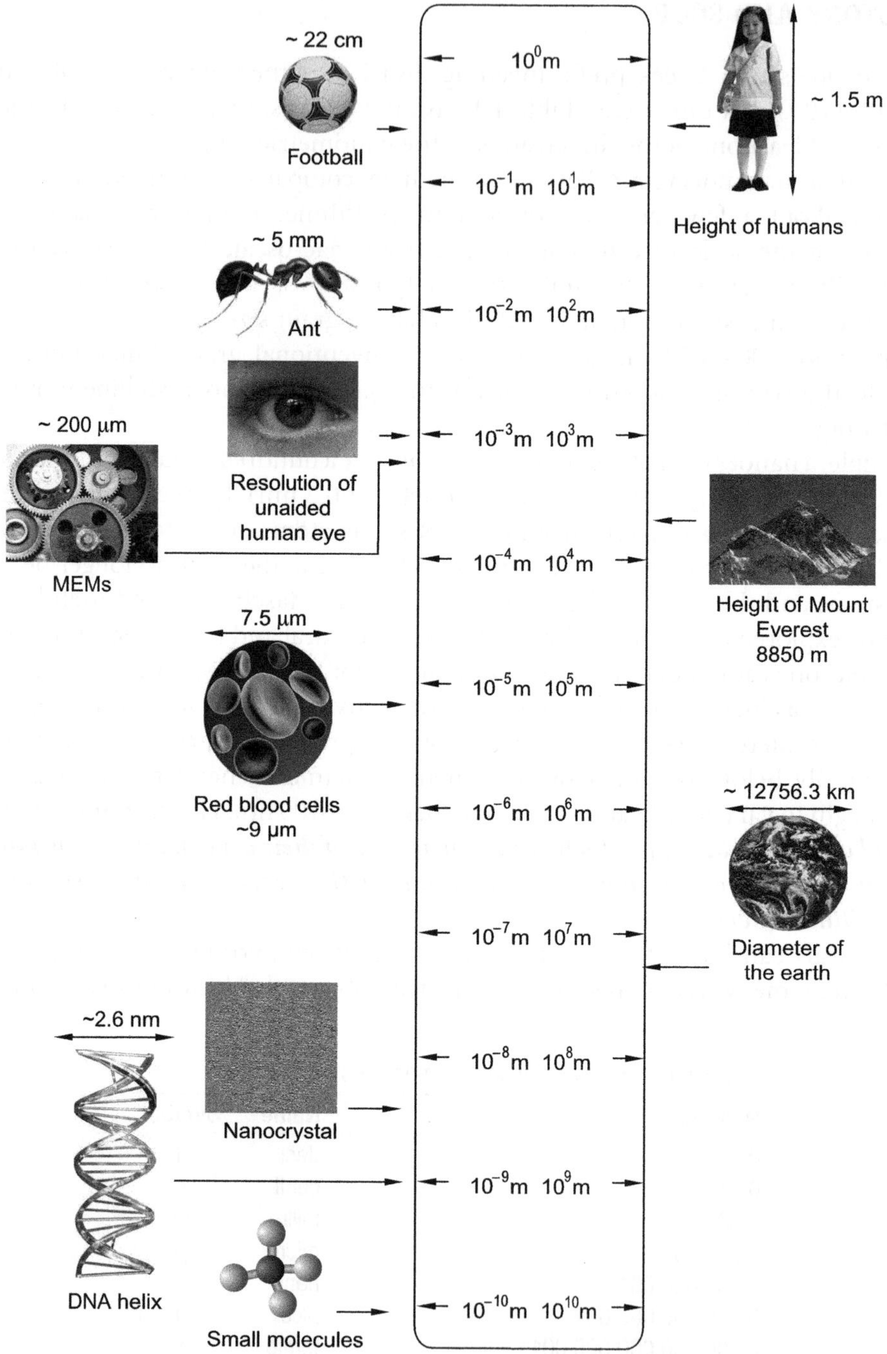

Fig. 1.2 Objects in different scales: micro to nano. A number of biological features are on the nanoscale and hence a deeper understanding of biological systems may be the key to success in nanoscience and nanotechnology.

CLUSTERS AND MAGIC NUMBERS

When atoms come together, they initially form two-dimensional clusters. When more atoms join the cluster, they become three-dimensional with topographically close-packed arrangements, before taking up a crystal structure. It has been found that clusters of certain critical sizes, i.e., clusters with a certain number of atoms in the group, are more stable than others. Such clusters are stabilized either by geometric or electronic considerations. **Magic number** is defined as the number of atoms in the clusters of critical sizes with higher stability. This effect was initially observed in gaseous metal atoms, in the early 1980s. Subsequently, a similar effect was also noticed during the condensation of atoms from a vapour phase on a substrate surface, for example, during thin film deposition.

Krypton atoms cluster together in stable sizes of 13, 55, 147, 309, 561,... atoms.

Magic numbers based on electronic shells were first observed in mass spectra of alkali metal clusters. The stability of such clusters as a function of the size did not follow a continuous function. It was seen that for some specific number of atoms in the cluster (*N*), *N* = 2, 8, 20, 40, 58, 92, etc., the free energy is lower, resulting in stabilization of the cluster.

Small clusters of atoms behave quite differently from bulk materials. Studies indicate that bulk properties are scalable only up to a certain critical grain size and below this limit, the behaviour of small clusters of atoms cannot be predicted on the same mechanistic or heuristic basis. Recent work has shown that atomic clusters containing certain magic numbers of atoms have quite distinct behaviour. For example, when 60 carbon atoms come together, the structure of carbon has unique icosahedral symmetry (crystal possessing 20 regular triangular faces) and such clusters are now widely referred to as fullerenes and have exceptional properties.

Krypton, being a noble gas, does not form strong chemical bonds with other atoms as well as with itself. However, Lethbridge and Stace of University of Sussex have succeeded in synthesizing clusters of krypton atoms with icosohedral symmetry. This was possible by allowing sudden expansion of krypton gas through a small orifice in a vacuum chamber. So far, clusters of 147 and 309 atoms have been detected with mass spectrography. There are models to suggest that the magic numbers of 13, 55, 147, 309, 561and 923 would have higher stability.

in the realm of a few hundreds of nanometres. The term 'nanotechnology' was first coined by Norio Taniguchi in 1974 to describe semiconductor processes such as thin film deposition and ion beam milling, where the features can be controlled at the nanometric level. The most widely accepted definition of nanotechnology to date appears on the NASA website:

> *...the creation of functional materials, devices and systems through control of matter on the nanometer length scale (1–100 nm), and exploitation of novel phenomena and properties (physical, chemical, biological) at that length scale.*

1.1.1 Nanomaterials are not new

Nanomaterials have been produced and used by humans for hundreds of years. However, the understanding of certain materials as nanostructured materials is relatively recent, made possible by the advent of advanced tools that are capable of resolving information at nanoscale.

- We now know that the beautiful ruby red colour of some ancient glass paintings is due to gold and silver nanoparticles trapped in the glass matrix (Fig. 1.3; see Plate 1).
- The decorative glaze or metallic film known as 'luster', found on some medieval pottery, contains metallic spherical nanoparticles dispersed in a complex way in the glaze, which gives rise to special optical properties. The techniques used to produce these materials were a closely guarded secret, and are not completely understood even now.
- Carbon black is a nanostructured material that is used in car tyres to increase the life of the tyre and impart black colour. This material was discovered in the 1900s. Fumed silica, a component of silicon rubber, coatings, sealants and adhesives, is also a nanostructured material. It became commercially available in the 1940s. Steel (an alloy of iron and carbon) is believed to have been first prepared in India about 1500 years ago and is popularly known as wootz (*India's Legendary 'Wootz' Steel: An Advanced Material of the Ancient World* by Sharada Srinivasan and Srinivasa Ranganathan is worth reading). This steel was used to make swords, which were so strong and sharp that they could easily cut a helmet into two pieces. Very recently, high-resolution electron microscopy of such a steel (picked up from a museum) showed the presence of carbon nanotubes in them, which has surprised scientists. People now believe that the high strength of these steels may be due to the presence of these carbon nanotubes, which are known for their exceptionally large Young's modulus.

The development of advanced microscopic analysis techniques, including transmission electron microscope (TEM), atomic force microscope (AFM), etc., have given great impetus to the identification and characterization of nanomaterials and the correlation of their behaviour with their structure. Figure 1.4 (see Plate 1) shows how the nano-world can be visualized using an AFM.

1.1.2 Early applications of nanotechnology: Nano-gold

Nanotechnology does not pertain to just miniaturization but goes beyond that. Materials in the nanometre-scale exhibit uniquely different physical, chemical and mechanical properties compared to bulk materials. Gold, for example, under ordinary conditions is a yellow, inert metal, capable of conducting electricity. If a centimetre-long gold foil is taken and broken into a dozen equal pieces, the pieces will still appear golden yellow. However, when the pieces are broken down about a million times, into bits just a few nanometres wide, almost every characteristic changes. Nano-gold no longer glitters with a golden yellow metallic lustre. Reflected light of gold nanoparticles varies in colour, depending upon their dimensions. Particles that are about 50 nm in diameter appear blue or purple, at 25 nm they are red, and at 1 nm they are orange (Fig. 1.5; see Plate 2).

Figure 1.6 (see Plate 2) is an excellent illustration of how the varying colours of nano-gold here used in ancient days for making tinted glass and the Roman Lycurgus cup. Could you have imagined that a noble material like gold could actually become highly reactive and be used as a powerful catalyst when synthesized in nanometric size? The melting point of gold also drops by almost 50% when the grain size is reduced below 10 nm. Similar changes have been observed in different nanomaterials with regard to a variety of other properties like magnetism and conductivity.

1.1.3 Publications on nanoscience and nanotechnology

In the past 10 years, the number of groups and laboratories engaged in the study of fundamental science, engineering and applications of nanostructured materials has grown almost exponentially. The significant growth in the number of publications in this area is a testimony to this fact (Fig. 1.7).

Many international and regional conferences or symposia have been held on nanostructured materials in the last two decades. The first international conference on nanostructured materials (Nano 1992) was held in Cancun, Mexico, in 1992. Subsequently, the conference was organized biannually in different parts of the world [Stuttgart, Germany (1994), Kona, Hawaii, USA (1996), Stockholm, Sweden (1998), Sendai, Japan (2000), Orlando, USA (2002), Wiesbaden, Germany (2004), Bangalore, India (2006), Rio, Brazil (2008) and Rome, Italy (2010)].

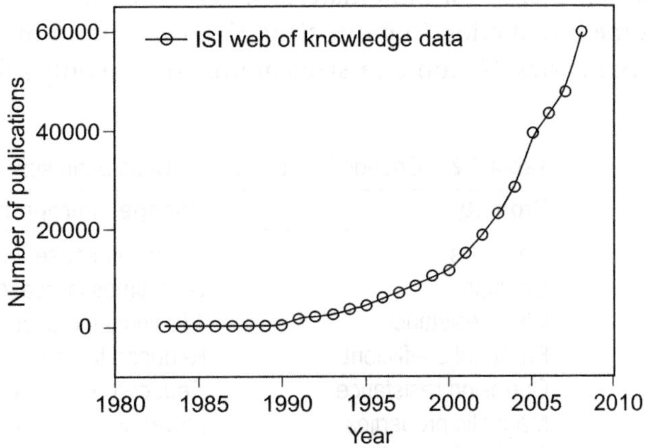

Fig. 1.7 Number of publications in the field of nanotechnology.

In India the Government has constituted the Nano Science and Technology Initiative (NSTI) (which is currently referred to as the Nano Mission programme) as a thrust activity of the Department of Science and Technology (DST), and has already supported research in this field to the tune of Rs. 200 crores over the last five years. In the current five year plan, this has been enhanced by about five times. NSTI has also initiated a conference series named ICONSAT (International Conference on Nano Science and Technology) for exchange of ideas by researchers in this multi-disciplinary field. The first ICONSAT was held at Kolkata in 2003, the second one at New Delhi in 2006, the third one at Chennai in 2008 and the fourth one at Mumbai in 2010.

1.2 CAN SMALL THINGS MAKE A BIG DIFFERENCE?

Why is the study of nanomaterials gaining wide importance and increased scientific attention? So what if the dimensions of a material are in the nanometric scale? Can small things make a big difference? Surprisingly, the answer is an emphatic YES. The modifications in the properties due to reduction in grain size to nanoscale dimensions are very large, and in most cases the resultant properties are superior to those of conventional materials. It is no wonder that nanomaterials are finding use in a large number of applications. More and more potential applications of nanomaterials are being discovered. For example, the change in properties of Ni when it is made in nanocrystalline form is shown in Table 1.2. It should be recognized, however, that there are secondary effects on properties, since commercially pure Ni contains impurity atoms that would prefer to segregate to the boundaries between grains. The higher concentration of impurity atoms at the grain boundaries will alter the bulk properties of solids.

1.2.1 Nanosize and properties

Figure 1.8 shows that nearly all properties like hardness, strength, ductility, elastic modulus, melting point, density, thermal conductivity, thermal expansion coefficient, diffusivity, and so on, change for nanomaterials. Why should the material behaviour vary so significantly by a mere reduction in grain size? Nanostructured materials are composed of grains and grain boundaries. Nanometre-sized grains contain only a few thousands of atoms within each grain.

Table 1.2 Change in properties of Ni as grain size is changed from 10 μm to 10 nm

Property	Change in property in comparison to bulk
Hardness	5 times increase
Strength	3–10 times increase
Wear resistance	170 times increase
Frictional coefficient	Reduced to half
Corrosion resistance	Reduced or localised corrosion is stopped
Magnetic properties	Lower coercivity, saturation magnetisation reduced by 5%
Electrical properties	Resistivity increased by 3 times
Hydrogen diffusion	Higher
Electrocatalytic properties	Improved electrocatalytic activities for hydrogen evolution

A large number of atoms reside at the grain boundaries, as shown in Fig. 1.9. As the grain size decreases, there is a significant increase in the volume fraction of grain boundaries or interfaces and triple junctions, as shown in Fig. 1.10. With increase in defect density, or in other words, when the fraction of atoms residing at defect cores like dislocations, grain boundaries and triple junctions becomes comparable with that residing in the core, the properties of the material are bound to be governed to a large extent by defect configurations, dynamics and interactions. For example, the way in which a crack grows in a larger-scale bulk material is likely to be different from the same in a nanomaterial where crack and particle size are comparable. Hence the mechanical and chemical properties of nanomaterials are significantly altered due to defect dynamics. The elastic modulus of nanomaterials can be significantly different from that of bulk alloys, due to the presence of increased fraction of defects. Nanocrystalline ceramics are tougher and stronger than those with coarse grains. Nano-sized metals exhibit significant increase in yield strength and the toughness decreases. It has also been shown that electrical, optical and magnetic particles are influenced by the fine-grained structure of these materials. As the technical capability to tailor and modulate dimensions at the nanoscale has improved greatly, it has become possible to realise the fascinating properties of nanostructures.

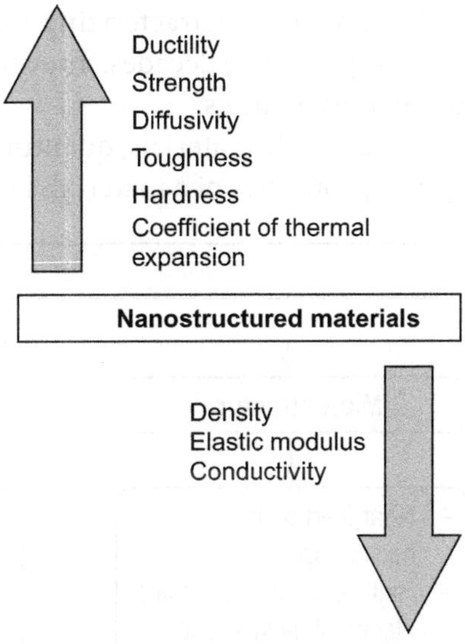

Fig. 1.8 Schematic diagram showing how different properties are affected in the nano-regime.

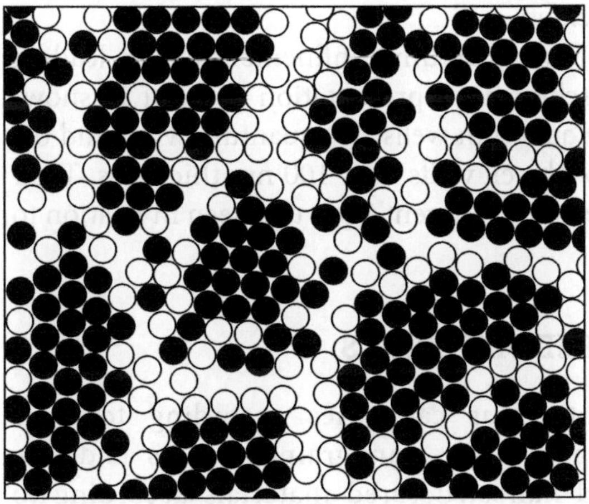

Fig. 1.9 The hypothetical structure of a nanomaterial. The black circles indicate atoms in the grain, while the white circles indicate atoms at the grain boundaries.

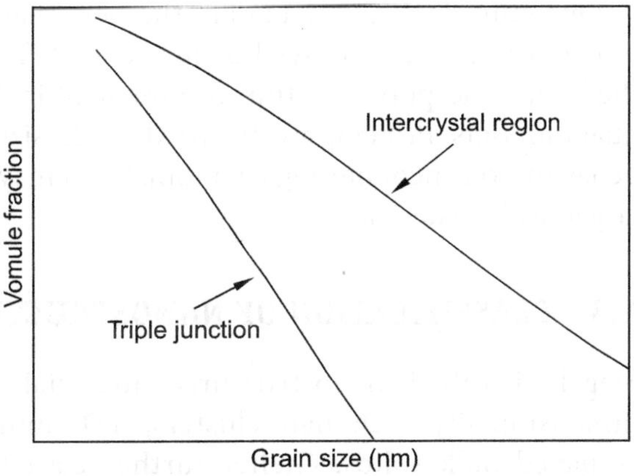

Fig. 1.10 Increase in the intercrystalline region (grain boundaries) and triple junctions with decrease in grain size of nanomaterials.

A number of characterizing tools (shown in the illustration below) have been developed over the past three decades. They have helped in understanding the behaviour of nanomaterials and nanostructures.

In nanoscale materials, quantum confinement can also lead to different electromagnetic and optical properties of a material. Due to the Gibbs–Thomson effect (see Chapter 2), the melting

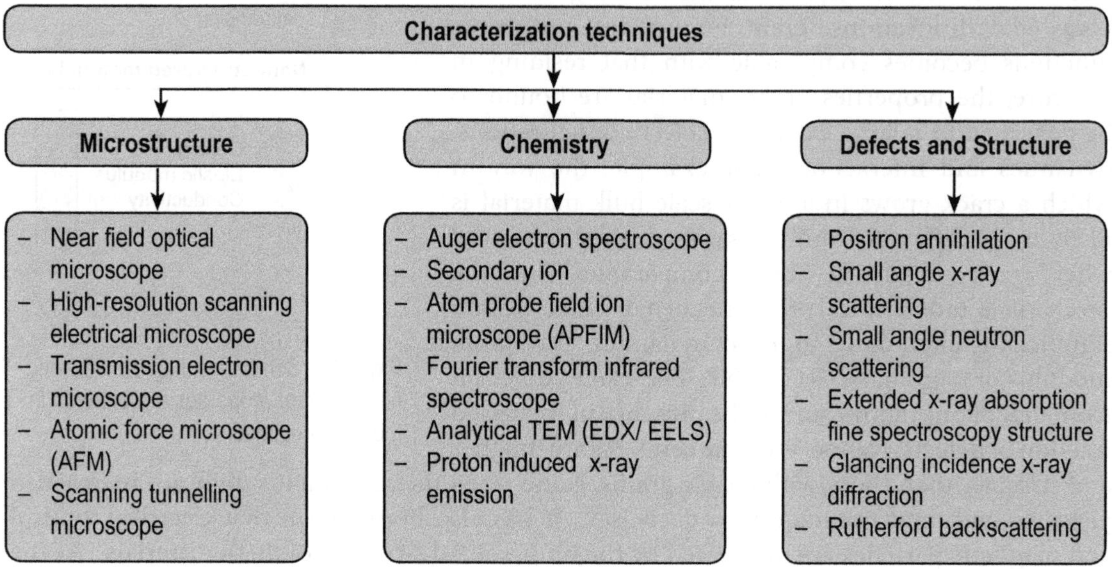

point of a free-standing particle is lowered if it is a few nanometres in size. Due to quantum mechanical forces that are exhibited at these length scales, nanomaterials may become better electrical conductors, be able to transfer heat better, etc.

Quantum dots (Fig. 1.11) are the best example of quantum confinement effects leading to bandgap tuning, etc (see box for details). The schematic representation in Fig. 1.12 shows the basic concept of quantum confinement in 2D quantum wells, 1D quantum wires and 0D quantum dots. Proteins are 10–1000 nm in size, with cell walls of 1–100 nm thickness. Their behaviour on encountering a nanomaterial may be quite different from that seen in relation to larger-scale materials.

1.3 CLASSIFICATION OF NANOSTRUCTURED MATERIALS

Siegel classified nanostructured materials into four categories according to their dimensionality: 0D: nanoclusters, 1D: multilayers, 2D: nanograined layers and 3D: equiaxed bulk solids. Gleiter further classified nanostructured materials according to the composition, morphology and distribution of the nanocrystalline component. This classification includes many possible permutations of materials and is quite broad. According to the shape of the crystallites, three categories of nanostructured materials are distinguished (Fig. 1.13):

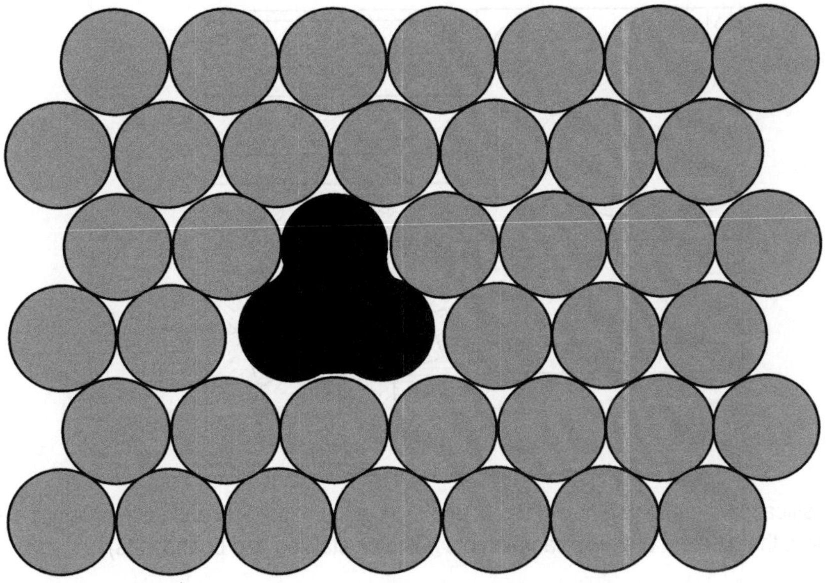

Fig. 1.11 Schematic of a quantum dot in a matrix. The cluster of atoms shown in black are embedded in the matrix of atoms in grey. This is a schematic representation of a quantum dot. (*Source:* http://en.wikipedia.org/wiki/File:Matrix_Isolated.JPG).

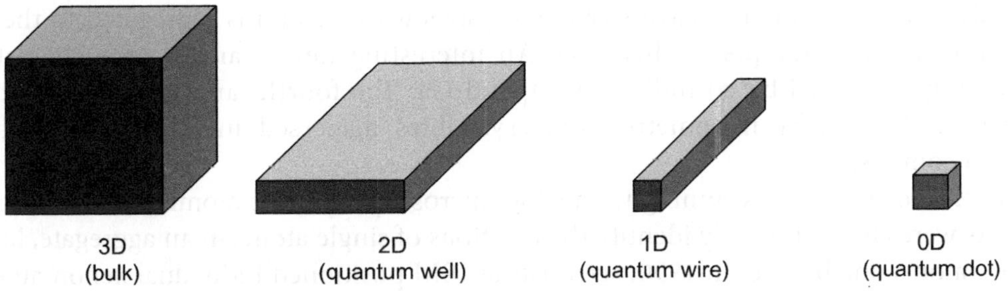

Fig. 1.12 Schematic diagram of quantum confinement in quantum well, quantum wire and quantum dot.

- Rod-shaped crystallites (with layer thickness or rod diameter of the order of a few nanometres)
- Layer-shaped crystallites
- Nanostructures composed of equiaxed nanometre-sized crystallites.

Depending on the chemical composition of the crystallites, the three categories of nanostructured materials may be grouped into four families. The simplest case is where all crystallites and interfacial regions have the same chemical composition. Examples of this family are semicrystalline polymers or nanostructured materials made up of equiaxed nanometre-sized crystals.

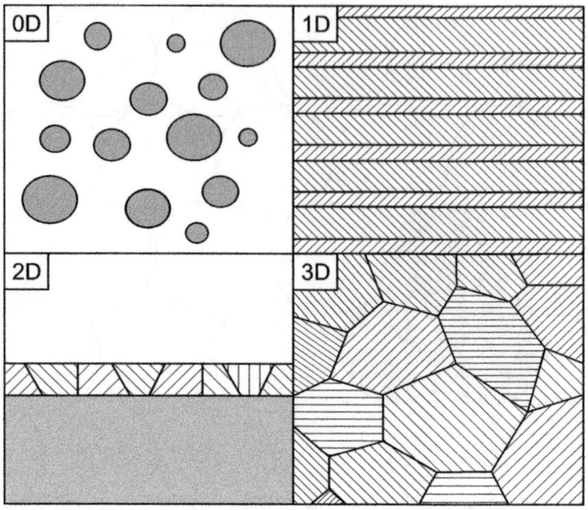

Fig. 1.13 Classification of nanomaterials. (The boundary regions of the first and second family are indicated in black to emphasise the different atomic arrangements in the crystallites and boundaries).

Nanostructured materials belonging to the second family consist of crystallites with different chemical compositions. Quantum well structures are well-known examples of this case. If the compositional variation occurs primarily between the crystallites and the interfacial regions, the third family of nanostructured materials is obtained. Nanostructured materials consisting of nanometre-sized W crystals with Ga atoms segregated to the grain boundaries are an example of this type. An interesting new example of such materials was recently produced by co-milling Al_2O_3 and Ga. The fourth family of nanostructured materials is formed by nanometre-sized crystallites dispersed in a matrix of different chemical composition.

With the advent of the scanning tunnelling microscope and the atomic force microscope, scientists were able to not only identify the positions of single atoms in an aggregate, but also to move atoms one by one. In 1990, scientists at IBM positioned individual xenon atoms on a nickel surface, using a scanning tunnelling microscopy probe. Subsequently, atomic scale manipulation has been demonstrated by several researchers for scribing characters at atomic scale (Fig. 1.14). In the mid-1980s, a new class of materials hollow carbon spheres—was discovered. These spheres were called bucky balls or fullerenes, in honour of the architect and futurist Buckminster Fuller. Fuller designed a geodesic dome with geometry similar to that found at the molecular level in fullerenes. The C_{60} (60 carbon atoms chemically bonded together in a ball-shaped molecule, Fig. 1.15) buckyballs inspired research that led to the fabrication of carbon nanofibres, with diameters under 100 nm. In 1991, Iijima of NEC in Japan reported the first observation of carbon nanotubes, which are now produced by a number of companies in commercial quantities. It may be noted that images of carbon nanotubes were published by Iijima several years prior to the discovery of fullerenes in 1985 by Robert Carl, Harold Kroto and Richard Smalley.

QUANTUM CONFINEMENT: QUANTUM DOTS

Quantum dots are clusters of atoms of a semiconductor material. The electrons in quantum dots have a range of energies. Quantum concepts such as energy levels, bandgap, conduction band and valence band apply to quantum dots too. However, the difference is the following. Excitons have an average physical separation between the electron and hole, known as the exciton Bohr radius, which is different for each material. In bulk, the dimensions of the semiconductor crystal are much larger than the exciton Bohr radius, allowing the exciton to extend to its natural limit. However, when the size of a semiconductor crystal approaches the material's exciton Bohr radius, then the electron energy levels can no longer be treated as continuous and must be treated as discrete, which means that there is a small and finite separation between energy levels. This phenomenon is called *quantum confinement* and the semiconductor material ceases to resemble bulk, and can instead be called a quantum dot. This has significant influence on the absorptive and emissive behaviour of the semiconductor material.

Because the electron energy levels of quantum dots are discrete, the addition or subtraction of just a few atoms to the quantum dot has the effect of altering the boundaries of the bandgap. The bandgap in a quantum dot will always be energetically larger and hence the radiation from quantum dots is 'blue shifted' (shifted to lower wavelengths), reflecting the fact that electrons must fall a greater distance in terms of energy and thus produce radiation of a shorter wavelength. In quantum dots, the size of the bandgap is controlled by changing the size of the dot. As the emission frequency of a dot is dependent on the bandgap, it is possible to control the output wavelength of a dot with high precision. Thus, it is possible to tune the bandgap of a dot, and thereby specify its 'colour' matching with the requirements of an application.

1.4 FASCINATING NANOSTRUCTURES

Nanostructured materials may occur in several different geometric configurations including wires, tubes, rods, horns, shells, pores, etc. They possess unique properties and are being developed for specific applications. Some of these interesting and emerging trends in nanostructures are described below.

Nanowires These can be defined as 1D nanostructures with nanometric width dimensions and exhibiting aspect ratios (the ratio between length and width) of 1000 or more (Fig. 1.16). Nanowires exhibit interesting properties deviating from bulk behaviour, due to quantum confinement in the lateral dimension. Nanowires are also, therefore, frequently referred to as

Fig. 1.14 Scanning tunnelling microscope (STM) image of PTCDA molecules adsorbed on a graphite surface. The logo of the Center for NanoScience (CeNS) in Munich, Germany, was written with the STM tip by reducing the tip–sample distance (compared to the imaging distance) and moving the tip along predefined vectors (nanomanipulation paths). This action removed molecules from the adsorbate layer along the manipulation paths. Line width of the STM-written logo: 1–3 nm. Image processing software: SPIP; 3D enhancement was applied. (*Source:* http://en.wikipedia.org/wiki/File:Cens_nanomanipulation3d_Trixler.jpg).

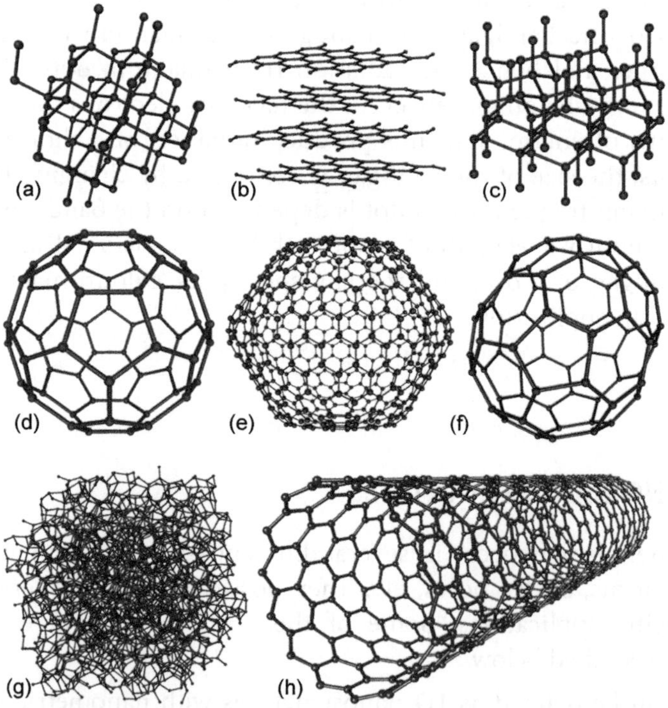

Fig. 1.15 Allotropes of carbon: (a) Diamond, (b) Graphite, (c) Lonsdaleite, (d) C60 (Buckminsterfullerene), (e) C540 (see Fullerene), (f) C70 (see Fullerene), (g) Amorphous carbon, (h) Single-walled carbon nanotube. (Source: http://commons.wikimedia.org/wiki/File: Eight_Allotropes_of_Carbon.png).

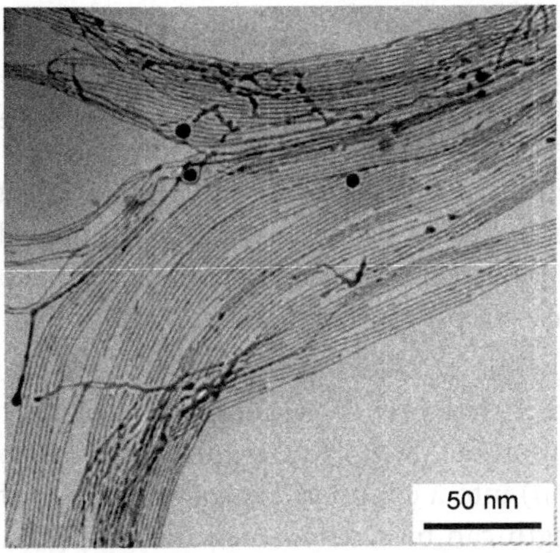

Fig. 1.16 Molecular scale Au nanowires (< 2 nm diameter) produced by wet chemical synthesis. The wires are formed by oriented attachment of amine-capped Au nanoparticles in a toluene medium. (*Courtesy:* N. Ravisankar, IISc, Bangalore).

'quantum wires'. So far, nanowires of several metals (Au, Ni, Pt), semiconductors (InP, Si, GaN) and insulators (SiO_2, TiO_2) have been fabricated. Nanowires can find interesting applications in the field of electronic, opto-electronic and nano-electromechanical sensors and devices. They are also used as metallic interconnects in quantum devices and for toughening advanced composites. It is possible to alter the growth parameters of nanowires so that they grow in a helical formation. Such nanowires are referred to as nanosprings. Nanosprings could have many applications as mechanical components with miniaturization of devices.

Nanorods These differ from nanowires in their aspect ratio. Standard aspect ratios for nanorods are 3–5. This means that unlike nanowires with an unconstrained dimension along the longitude, nanorods have all their dimensions in the range 1–100 nm and hence are 3D nanostructures. They may be synthesized from a variety of materials and can find diverse applications ranging from display technologies (the reflectivity of the rods can be changed by changing their orientation with an applied electric field) to micromechanical switches.

Nanoshells These are those structures where the nanocrystalline particles are coated with a thin layer of a different material with thickness in nanometric dimensions. Due to their nanometric width and size, nanoshells can exhibit interesting quantum confinement effects. For example, gold nanoshells coated on dielectric silica nanoparticles have been widely studied in drug research. It is possible to vary the light absorption and emission characteristics of gold nanoshells by controlling the thickness and size of the nanoparticles (particularly as they approach the wavelength dimensions of the light medium used). Study of plasmon resonance wavelength shift as a function of nanoshell composition for the case of a gold/silica nanoshell with a 100-nm core, demonstrates the ability to tune the optical resonance of nanoshells to a desired wavelength. This property is critical in nano-drug delivery for *in vivo* therapeutic applications.

Laser light is not strongly absorbed generally by human tissues and blood. However, due to resonance effects, it is possible for nanoshells to absorb significant amounts of incident laser light, resulting in intensive localized energy absorption. By attaching certain antibodies to such nanostructures, it is possible to provide site specificity, for example, for selective segregation of such nanoparticles in cancerous or tumour cells. Subsequently, as the body is irradiated with laser light, there will be selective heating of nanoshells concentrated at cancerous cells, leading to selective burn out of such deleterious cells. Nanoshells can also find application in nano-medical diagnostic tools, for example, in thermal imaging.

Nanotubes These are tubes with diameters in the nanoscale. Although nanotubes of various other materials have been reported, carbon nanotubes are by far the most important group. A single-walled carbon nanotube (SWCNT) is obtained by rolling a sheet of graphite (a hexagonal lattice of carbon) into a cylinder. Typical diameters of SWCNTs are in the range of 0.7–1.4 nm and their length can be several micrometres. Multi-walled carbon nanotubes (MWCNTs) can be regarded as a coaxial assembly of SWCNTs (Fig. 1.17, see Plate 3). MWCNTs have their diameters in the range of 5–50 nm. Carbon nanotubes are unique nanostructures with remarkable electronic and mechanical properties, some of which are due to the close relation between the carbon nanotubes and graphite, and some from their one-dimensional aspects. Carbon nanotubes also exhibit good nano-mechanical properties due to their high Young's modulus. The modulus of single-walled (1210 GPa) and multi-walled (1260 GPa) carbon nanotubes is now established as a world record as the highest known modulus for any material. In comparison, metallic wires such as copper and steel have a modulus of only 110 and 200 GPa, respectively.

Nanofluids These are fluids (e.g. water, ethylene glycol, lubricants) with dispersions of nano-sized particles (e.g. carbon, metals, metal oxides, etc.). The nanofluids are characterized by unique properties and have several applications. The most popular nanofluids are magnetic nanofluids (ferrofluids) and thermal nanofluids. In case of thermal nanofluids, improvements in energy efficiency and convective heat transfer have been demonstrated by many investigators and thermal conductivity of the fluids has been enhanced to as high as 150% with the dispersion of minute quantities of nanoparticles. A number of applications such as drug delivery, magnetic storage media, refrigerant chillers, electronic manufacturing, cosmetics, pharmaceuticals, power generation, air-conditioning, etc. are envisaged for nanofluids.

Nano-fantasies

Nano-guitar It is the world's smallest guitar and is 10 μm long (about the size of a single cell) with six strings, each about 50 nm wide. Researchers made this device at the Cornell Nanofabrication Facility, bringing micro-electromechanical devices (MEMS) to an even smaller scale—the nano-sized world.

Nano-balance It is small enough to weigh viruses and other submicron-scale particles. It works on the principle of variation in resonance frequency of the nanotube by changing its mass. By knowing the resonance frequency of pure nanotubes, the mass of attached particles can be studied. A nano-balance could be useful for determining the mass of objects on the femtogram to picogram range. Such a nano-balance has been developed at the Georgia Institute of Technology, Atlanta, USA, by growing long individual filaments of carbon nanotubes.

1.5 APPLICATIONS OF NANOMATERIALS

Since nanomaterials possess unique chemical, physical and mechanical properties, they can be used for a wide variety of applications, ranging from toothpaste to satellites. Nanotechnology is finding application in virtually all fields ranging from science to engineering, influencing our lifestyle in many ways. The consumer world is being flooded with 'nanotechnology enhanced' products, with many more to come. Nanomaterials are finding application in cosmetics, textiles, healthcare, tissue engineering, catalysis, functional coatings, medical diagnosis and therapeutics, sensors and communication engineering, water and air pollution treatment. Some of these applications are discussed below.

Electronic devices Numerous nano-electronic applications are in use for communication and computing purposes. The days of massive computer stations occupying an entire room with huge punch cards to process each program are behind us. Today's multi-functional laptops and palmtops are more user friendly, faster, handy and have large memory capacities. Mobile phones, pocket-sized memory storage devices and the widely used MP3 players, iPods and iPads are perhaps the most convincing benefits of nanotechnology. All this has been possible due to the shrinking sizes of electronic devices enabled by nanotechnology.

Miniaturization is the *mantra* of the electronics industry, which is driving research on reduction in the size of transistors, resistors, capacitors, etc. *Moore's Law,* an empirical observation made in 1965, states that the number of transistors on an integrated circuit for a minimum component cost doubles every 24 months (Fig. 1.18). Even four decades later, the growing electronics industry still follows this law. The microprocessors containing these smaller components can work at enhanced speed, thereby enabling computations at far greater speeds than what was possible a few years back. However, there are several technological difficulties to overcome, such as lack of ultrafine precursors to manufacture these components, poor dissipation of heat generated by these microprocessors, etc. Nanomaterials can help the industry overcome these barriers.

Nanostructures hold promise for the development of non-volatile, radiation hard, high-density (terabit/cm^2) memory with nanosecond read/write times; there are several technologies being explored. While one terabit per square centimetre sounds astronomically large, a single movie, *The Lord of the Rings,* required 27 terabytes (about 200 terabits) for the first instalment alone.

Researchers at IBM Zurich have been working on a novel memory device based on a mechanical storage method. This system, called 'Millipede', uses an array of microcantilevers to create a pattern of nanoindentations in a polymer medium. This method has demonstrated storage densities as large as 1Tb/in^2. Researchers at Nantero, USA, and Korea are developing

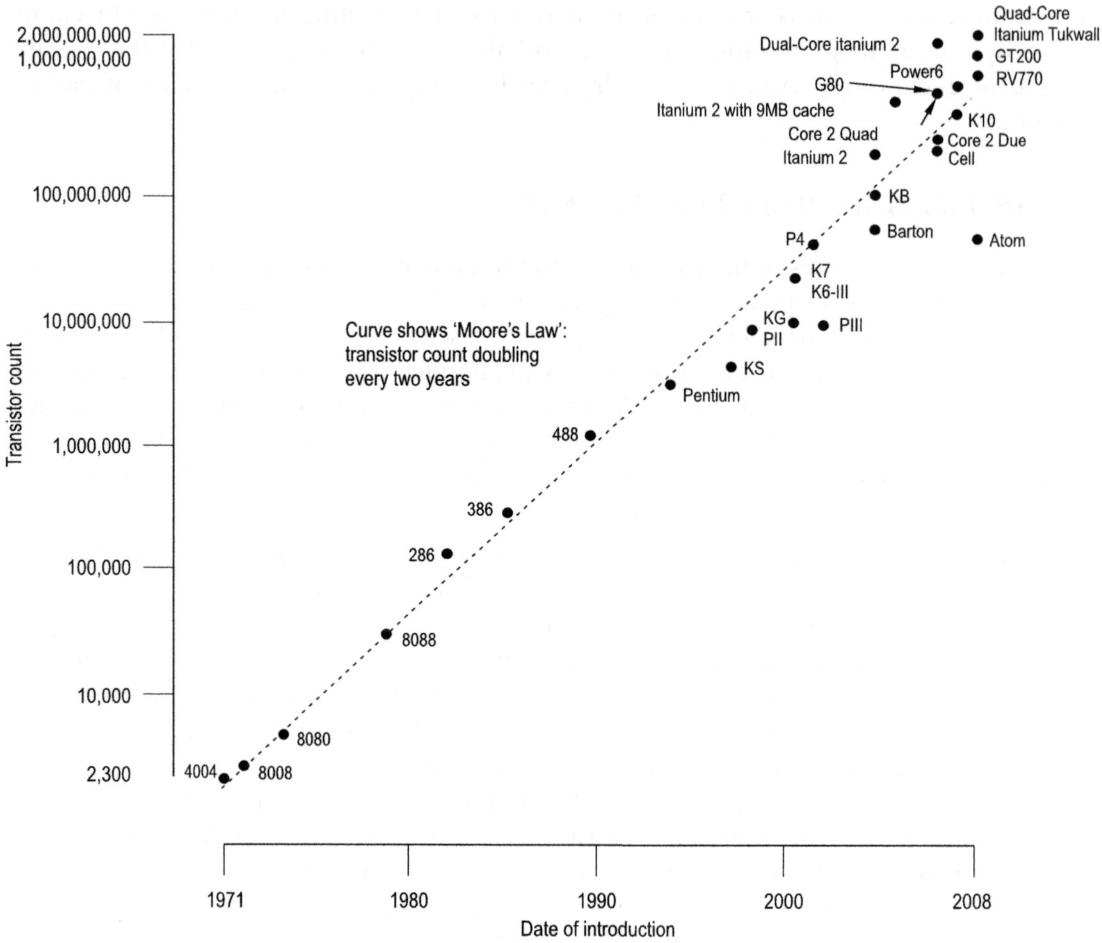

Fig. 1.18 Moore's Law: The number of transistors in an integrated circuit for minimum component cost doubles every 24 months. (*Source:* http://commons.wikimedia.org/wiki/File:Transistor_Count_and_Moore%27s_Law_-_2008.svg).

carbon-nanotube based memory concepts. The giant magnetoresistance (GMR) effect has led to a strong increase in the data storage density of hard disks and made the gigabyte range possible. The so-called tunnelling magnetoresistance (TMR) is very similar to GMR and is based on the spin dependent tunnelling of electrons through adjacent ferromagnetic layers. Both GMR and TMR effects can be used to create a non-volatile memory for computers, such as the magnetic random access memory. The company Freescale (as well as IBM) has a GMR memory device on the market with 4 Mbits, while researchers at Naval Research Laboratory (NRL), USA, are developing the next generation GMR memory that may reach terabit densities.

Opto-electronic devices Opto-electronic devices convert electricity to light and vice versa. They have broad bandwidth and efficiency, and find application in LEDs (light emitting diodes), OLEDs (organic LEDs), LCDs (liquid crystal displays), laser diodes, modulators, CMOS (complementary metal oxide semiconductor) and CCD (charge-coupled device) photodetectors, and solar cells. Opto-electronic devices coupled with optical fibres have been extensively used in the fabrication of TFT (thin film transistor)-LCD laptop PC screens, automobile illuminations, mobile phone backlighting, VCD/DVD players, telecommunications and data communications (broadband communications), biotechnology (BioPhotonics), and digital cameras.

Quantum computers Currently, lasers produced from nanoscale quantum dots are being fabricated. The wavelength of the laser is reported to depend on the diameter of the quantum dot. Quantum dot lasers are cheaper and offer higher beam quality than conventional laser diodes. With the ability to synthesize quantum dots on a commercial scale, it has become possible to exploit the laws of quantum mechanics for novel quantum computers, using fast quantum algorithms. Quantum computers can perform several computations at the same time and are much faster. This class of computers would be useful to solve specific problems and supplement digital computation.

Insulation Aerogels are nanomaterials synthesized by the sol-gel process, which are porous, foam-like and extremely light-weight, and yet can withstand about 100 times their weight. They are currently being used for insulation in offices, homes, etc. They are also being used in 'smart' windows, which darken on a sunny day and lighten on a cloudy day.

Phosphors The use of nanophosphors such as zinc selenide, zinc sulphide, cadmium sulphide and lead telluride is expected to make high-definition televisions (HDTVs), personal computers and laptops affordable by an average household. Carbon nanotubes have been used for the production of displays with low energy consumption, primarily due to their higher efficiency field emission properties.

Cutting tools Microdrills (drill bits with diameter less than the thickness of human hair), used in the miniaturization of microelectronic circuits, are required to have enhanced wear resistance. Nanocrystalline carbides and nitrides are harder and wear-resistant, and hence are currently being used in these microdrills.

Medicine In the field of medicine, nanotechnology finds application in diagnosis, therapeutics, prosthesis materials and tissue engineering. Nanomaterials have dimensions similar to those of biological molecules and hence they are useful for biomedical applications. By attaching different biomolecules to nanomaterials, they can be used in medical applications for specific functions. Nanotechnology is being developed for both therapeutic (using nano-drug delivery systems) and diagnostic (nano-biosensors) applications.

- Gold nanoparticles attached to short segments of DNA can be used to detect the genetic sequence of a sample.
- Bandages embedded with silver nanoparticles are gaining popularity for providing anti-microbial protection and to aid faster wound healing.
- Nanostructured synthetic skin is being used for several skin graft applications.

- Biocapsules are increasingly being used as substitutes for diabetic insulin, targeted drug delivery, biosensors, etc.
- Respiration monitors utilizing nanomaterials can be more sensitive than conventional monitors.

Early detection of a disease always enables quicker and more effective cure and recovery. Nanotechnology-based sensors and medications have come in handy for the early diagnosis of several diseases. Nanoparticles have shown potential for detecting viruses, pre-cancerous cells, etc. Targeted drug delivery systems can deliver drugs effectively and conveniently with increased patient compliance, extended product life and reduced healthcare costs.

Nanobots or nano-robots are becoming a reality. They are devices that can be injected into the bloodstream to cure diseases at specific locations and their motion is controlled by an external field. Nanotechnology is thus envisioned as being capable of changing the way we live. Drug delivery systems that rely on nanomaterials are used for the targeted delivery of compounds characterized by low oral bioavailability due to poor water solubility, permeability and provide for longer sustained and controlled release. The overall drug consumption and side effects can be lowered significantly by coating drugs on nanoparticles. This highly selective approach reduces cost and human suffering. For example, dendrimers and nanoporous materials can transport small molecules of drugs to the desired location. Another application is based on small electromechanical systems: nano-electromechanical systems (NEMS) are being investigated for the active release of drugs. The use of gold-coated shells for cancer therapy is in an advanced stage of research.

Renewable energy Perhaps the greatest challenge for society and humankind is how to meet our ever-increasing demand for energy security. Although the primary source of energy on earth is derived from the sun, the fundamental steps for energy conversion, such as the transfer of charge, chemical reactions, transformation of molecular structure, etc., occur at the nanoscale. Development of nanotechnology has the potential to revolutionize the approaches to energy production. Some of the promising new areas for the use of nanotechnology in this field are: use of nanomaterials to extract hydrogen from water, to harvest energy from the sun and biomass, to store energy as hydrogen fuel cells, batteries and capacitors. Nanomaterials are also being used as advanced catalysts for energy conversion. Nanomaterials will also impact efficient utilization of energy for industries such as transportation, power generation and utilization, water management and purification and environmental cleanup. A number of studies are underway to deploy nanotechnology to meet future energy needs.

Catalysis One of the most lucrative areas for a nanotechnologist is catalysis. Chemical catalysis benefits significantly from nanoparticles, due to the extremely large surface-to-volume ratio. It is easy to understand from geometrical principles that the surface area to volume ratio is inversely proportional to the size of a particle. A variety of chemical reactions take place on the surface of a catalyst, and hence the larger the surface area, the more active the catalyst. Nanoscale catalysts thus open the way for numerous process innovations to make many chemical processes more efficient. The application potential of nanoparticles in catalysis ranges from fuel cells to catalytic converters and photocatalytic devices. Nanoparticles can be used as catalysts in

- automobile catalytic converters for the removal of noxious and toxic gases such as carbon monoxide and nitrogen oxide
- power generation equipment to prevent environmental pollution arising from burning gasoline and coal

Filtration Nanochemistry also finds immense use in wastewater treatment and air purification devices. One class of filtration techniques is based on the use of membranes with appropriately-sized pores, through which the liquid is allowed to pass. Nanoporous membranes used in nanofiltration have extremely small pores of less than 10 nm. Nanofiltration is used mainly for the removal of ions or the separation of different fluids. Ultrafiltration is used to remove particles with size between 10 nm and 100 nm. Renal dialysis is an important application of ultrafiltration. Magnetic nanoparticles offer an effective and reliable method to remove heavy metal contaminants from wastewater by the use of magnetic separation techniques. Nanoscale particles increase the efficiency to adsorb contaminants.

Elimination of pollutants Nanoscale materials can be used effectively in soil/sediment and water remediation. For sites where refractory organic contamination has penetrated deeply into the soil, it would be cost effective to develop remediation technologies that do not require excavation. The use of zero-valent iron particles has been shown to be effective in penetrating the required distances, and oxidizing selected organic contaminants. Research is underway to examine the fate of those iron particles to make sure they do not cause ancillary problems.

Sensors Sensors made of nanocrystalline materials are extremely sensitive to a change in their environment. Some of the applications for sensors made of nanocrystalline materials are smoke detectors, ice detectors on aircraft wings and automobile engine performance sensors.

Food Food packaging can be improved by placing anti-microbial agents directly on the surface of the coated nanocomposite film. The incorporation of nanoscale clay particles in a polymer matrix can result in lower oxygen and water permeation with better recyclability. This can protect food from drying and spoilage (incurred by oxygen access). Silver has been used as an anti-microbial agent for centuries. Recently, nanoscale silver has been introduced in anti-microbial containers to keep food fresh for longer.

Consumer products Nanotechnology is impacting the field of consumer goods, by providing products with novel functions ranging from easy-to-clean to scratch-resistant coatings. The most prominent application of nanotechnology in the household is self-cleaning or 'easy-to-clean' ceramic or glass surfaces. Nano-ceramic particles have improved the smoothness and heat resistance of common household equipment such as the flat iron and cooking pans.

Sports In the field of sports, nanotechnology is promising to enhance the life and performance of sports gadgets. Nanotechnology has been employed to produce tennis balls that last longer, rackets that are stronger and bowling balls that are harder. Nano-ski wax is easier to apply and more effective than standard wax. Nanotechnology enhanced golf balls can correct their own flight path so that they fly straighter than conventional balls.

Textiles There are several applications of nanotechnology in textiles and fabrics, ranging from anti-microbial, hydrophobic and self-cleaning applications. When the fabric is mixed with a hydrophobic material, it repels water and is also stain resistant. In the monsoon season, for example, drying such clothes will be much easier, and these clothes will be much more comfortable to wear. Nanoparticles have been incorporated into products such as nylon, polypropylene and other polymers to impart long-term anti-microbial characteristics even under harsh environmental conditions or after extensive thermal cycling. Nano-socks with nano-silver dispersions are odourless and also possess anti-microbial properties. The demand for minimizing or eliminating microbial growth on a variety of textile-based substrates has increased in healthcare, home furnishings, filtration and apparel sectors, among others. Nano-cameras mixed with nano-displays create an 'invisibility coating', which is useful for making camouflage clothing for the military.

The above are only a few applications of nanomaterials. Many new applications are being discovered every day and many more are yet to be discovered.

1.6 NATURE: THE BEST NANOTECHNOLOGIST

Nature is perhaps the foremost inspiration for nanoscientists and nanotechnologists. Over millions of years, many nanoparticles and devices have been perfected by nature through the process of evolution. Mere observation of the natural phenomena around us can reveal new directions and insights in nearly all domains of science and technology, and particularly nanoscience and nanotechnology.

The cell membranes, and several other functional organelles within the biological cell of living beings, are in fact of nanometric size (Fig. 1.19). Proteins and enzymes together account for all the metabolic activities in the body—an excellent illustration of the power of nanostructures. We are all aware that life on earth could not have been sustained without the life-saving phenomenon of *photosynthesis* that enables plant species to convert solar energy into biochemical energy forms. Photosynthesis is enabled in plants by nanoscale molecular machinery, consisting of pigment molecules like chlorophyll arranged inside stacked structures called thylakoid disks housed inside micrometre-sized chloroplast cells.

Have you ever wondered at the engineering skills of a web-spinning spider (Fig. 1.20)? The web is made up of nanofibres that are light and insoluble in water, but in terms of specific strength, they are stronger than steel. These webs that can withstand environmental effects such as rain, wind and sunlight are a marvel of nature. The spider has sufficient supply of raw materials to spin the web over great distances relative to its body size. The speed with which the spider is able to cast its net in an organized fashion under ambient conditions is intriguing. Yet another example of a nanostructure developed by nature is the ability of a gecko to walk across a ceiling against gravity. The self-cleaning nature of a lotus leaf comes from the nanospikes present on its surface; it helps the water droplets and dust particles (Fig. 1.21; see Plate 3) roll off with ease. Water striders (Fig. 1.22) are able to walk on the surface of water without getting wet because of the nano-grooves present in the microhairs of their legs.

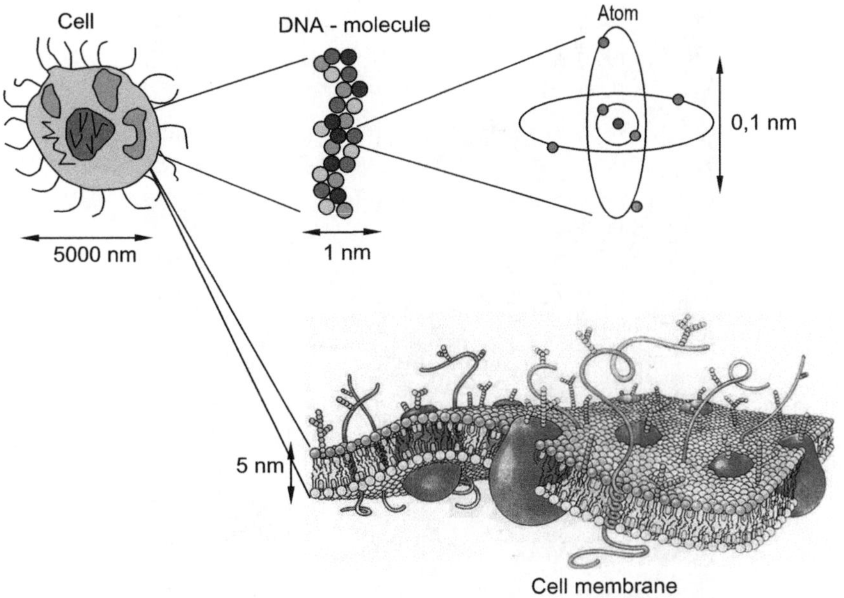

Fig. 1.19 Nanotechnology in nature. Biological features such as DNA, cells and membranes are of nanoscale.

Fig. 1.20 Spider's web: It is not only aesthetically pleasing but is also a biological wonder as the fibres of the web are the strongest known fibres—until the carbon nanotubes were discovered. (*Source:* http://commons.wikimedia.org/wiki/File: SpiderWeb.jpg).

Fig. 1.22 Water striders: Nano-grooves on microhairs are responsible for their ability to move on water without getting wet. (*Source:* http://en.wikipedia.org/wiki/File:Amenbo_06f5520sx.jpg).

We are familiar with the soft chalk used to write on blackboards. It is composed of loosely agglomerated calcium carbonate powder. Nature, however, has a method to make the same calcium carbonate agglomerates hard and tough—by synthesizing them in nanocrystalline form. The abalone, a mollusc, has a tough shell made of essentially the same stuff that is used to make chalk pieces. However, the nanocrystalline nature resists crack propagation and instead promotes crack deflection along the grain boundaries. Cracks have to essentially traverse larger cumulative length scales, which requires more energy and thereby enhances the toughness of the shells.

The sophisticated and efficient sensory systems in living species are perhaps some of the greatest scientific and engineering marvels perfected by nature by evolution over millions of years. They are also a knowledge resource and strong motivators for nanotechnologists attempting to develop smart sensors. Biological sensory systems are extremely sensitive and are only limited by quantum effects. The ability of our body to monitor body temperature and regulate it by means of controlled metabolic activity, with the help of nanomolecular proteins, is remarkable and intriguing. Scientists are already working on developing an electronic tongue and electronic nose, based on assembling a number of nanosensors, with each sensor having a specific role.

In the current age of communications, there is a continuous demand to enhance information storage capacity in electronic devices. Nature has perfected DNA, a nanomolecule, that not only stores but also manifests the entire personality of a species. Even more fascinating perhaps is that, nature synthesizes these nanomolecule more efficiently and quickly under ambient conditions. In contrast, all artificial nanomaterial synthesis methods adopted by us today are based on high-energy processes, usually involving high temperature, electrical or other forms of energy, high or very low pressures, etc. Hence, there is so much more to learn from nature, even in fundamental aspects of synthesis and functioning of nanomaterials.

1.7 CHALLENGES AND FUTURE PROSPECTS

With so many achievements already realised, it is perhaps pertinent to ask "Where is nanotechnology heading"? It is envisaged that nanotechnology will lead to tiny robotic devices, utilizing nanoelectronics, sensors and MEMS/NEMS for in-vivo monitoring and diagnosis of electro-optic deficiencies and malfunctions of human systems. Yet, the current applications of nanotechnology are much more mundane: stain-resistant trousers, better sun creams, tennis rackets reinforced with carbon nanotubes! There is a huge gap between what nanotechnology is believed to have promised and what it has actually delivered so far.

In his book, *Engines of Creation: The Coming Era of Nanotechnology*, published in 1986, Drexler imagined sophisticated nanoscale machines that could operate with atomic precision. He envisaged a particular way of achieving nanotechnology, which involved using hard materials like diamond to fabricate complex nanoscale structures by moving reactive molecular fragments into position. His approach was essentially mechanical, whereby tiny gears and bearings are integrated to make tiny robot factories, probes and vehicles.

Drexler postulated that since nanoscale machines are expected to be extensively employed in biological systems and would be synthesized in significantly large quantities under ambient conditions, it should be possible to discover the growth conditions to synthesize them for a variety of other applications as well. The beauty of nanotechnology is that it is truly multi-disciplinary, re-unifying the common threads between science, engineering and technology. It is so vivid, with possibilities left only to the constraint of perhaps human imagination. With a little exaggeration, it seems possible that materials with any desired physical, chemical or electronic properties can be tailor-made by playing with the nano-dimensions.

The next generation is going to be directly or indirectly exposed to a variety of nano-products ranging from cosmetics to sports, from medical to industrial, and also space applications. With the advent of any new revolutionary technology with enormous potential for applications, it is perhaps even more pertinent to assess the risks and challenges accompanying them. The effect of nanoparticles on biological and ecological systems in large is a subject to be studied with the highest priority. A new field, nano-toxicology, has evolved in order to probe this field. It is important to qualify the application of nanomaterials for industrial and large-scale societal applications, not only based on their properties but also based on their possible long-term side effects.

SUMMARY

- The unique properties of nanomaterials are due to the presence of a high concentration of defects.
- The word *'nano'* refers to a Greek prefix meaning dwarf or something very small. It depicts one billionth (10^{-9}) of a unit.
- Nanomaterials refer to the class of materials with at least one of their dimensions in the nanometric range. They can be metals, ceramics, polymers or composites.
- Nanomaterials exhibit uniquely different physical, chemical and mechanical properties compared to bulk materials.
- A number of characterizing tools have been developed over the past three decades and have helped in understanding the behaviour of nanomaterials and nanostructures.
- Nanostructured materials may occur in several different geometric configurations including wires, tubes, rods, horns, shells, pores, etc.
- Over millions of years, a multitude of nanoparticles and devices have been perfected by nature through the process of evolution.
- The cell membranes, and several other functional organelles within the biological cell of living beings are in fact of nanometric size.
- It is envisaged that nanotechnology will lead to tiny robotic devices, utilizing nano-electronics, sensors and MEMS for invivo monitoring and diagnosis of deficiencies and malfunctions of human systems.

Norio Taniguchi (1912–99) of Tokyo Science University coined the term nanotechnology in 1974 to describe semiconductor processes such as thin film deposition and ion beam milling, which exhibit control on the order of a nanometre. He defined nanotechnology as that which consists of the "processing of separation, consolidation, and deformation of materials by one atom or one molecule". The early work of Prof. Taniguchi was on abrasive mechanisms during high precision machining of hard and brittle materials. At Tokyo Science University, he pioneered the application of energy beam techniques to ultra precision materials processing, which included electro discharge, microwave, electron beam, photon (laser) and ion beams. Professor Taniguchi received Euspen's 1st Lifetime Achievement Award in 1999. The citation on Prof. Taniguchi's award reads: "In recognition of his unique and outstanding contributions to research and development in the ultra precision materials processing technologies and in 1974, being the first to formulate and use the term Nanotechnology. Through his vision, writings and example of total dedication to his field of endeavour he has stimulated the development of what will be one of the dominant technologies of the 21st Century."

CNR Rao is the President and Linus Pauling Research Professor of the Jawaharlal Nehru Centre for Advanced Scientific Research, Bangalore. He steers the Nano Science and Technology Initiative (NSTI) of the Department of Science and Technology, Government of India. Although Prof. Rao started his career working on spectra and structure of molecules, his research interests have slowly moved towards the synthezis and characterization of designer solids with novel structures and properties. Several nanoparticles and nanoparticle assemblies, open framework structures and porous solids have been synthesised by him exhibiting fascinating scientific behaviour. He has authored nearly 1400 research papers and edited or written 40 books in materials chemistry. The Einstein Gold Medal of UNESCO, Hughes Medal of the Royal Society, and Somiya Award of the International Union of Materials Research Societies (IUMRS), Dan David Prize for materials research from Israel and the first India Science Prize are among the recognitions that he has received.

Richard P Feynman (1918–88) was one of the greatest theoretical physicists of the twentieth century. This quote of his is regarded as the 'start' of nanotechnology, "The principles of physics, as far as I can see, do not speak against the possibility of maneuvering things atom by atom. It is not an attempt to violate any laws; it is something, in principle, that can be done; but in practice, it has not been done because we are too big."

K Eric Drexler is the pioneer of productive nanosystems. He has authored the book *Engines of Creation: The Coming Era of Nanotechnology*, which is possibly the first book in this field. He has also written another book *Nanosystems: Molecular Machinery, Manufacturing, and Computation*, which has also become very popular. He is currently Chief Technical Advisor of Nanorex, which is engaged in developing software for the design and simulation of molecular machine systems. He received his doctoral degree in the field of molecular nanotechnology from MIT in 1991, which is the first PhD degree in this field.

Sumio Iijima discovered carbon nanotubes in 1991. He has been a Senior Research Fellow at NEC Corporation since 1987 and a Professor of Materials Science at Meijo University in Nagoya. He has the leading role in the International Cooperative Research Project 'Nanotubulites' of Japan Science and Technology Corporation (ICORP/JST). He is also the Director of the Research Center for Advanced Carbon Materials of National Institute of Advanced Industrial Science and Technology (AIST).

Prominent individuals in nanoscience and technology

Richard P Feynman: First mentioned the concept of 'nano' in a 1959 lecture
Norio Taniguchi: Defined the term 'nanotechnology'
K Eric Drexler: Promoted technological significance
Sumio Iijima: Discovered nanotubes
Don Eigler: Studied manipulation of surface atoms and the effects on electron density of states
Richard Smalley: Co-discoverer of buckminsterfullerene
Harry Kroto: Co-discoverer of buckminsterfullerene
Erwin Müller: Invented the field ion microscope and the atom probe
Gerd Binnig: Co-inventor of the scanning tunnelling microscope
Heinrich Rohrer: Co-inventor of the scanning tunnelling microscope
Phaedon Avouris: Made the first electronic devices made of carbon nanotubes
Alex Zettl: Built the first molecular motor based on carbon nanotubes
PM Ajayan: Pioneer in nanotubes research
CNR Rao: Most prominent researcher and promoter of nanomaterials research in India
H Gleiter: First produced nanostrctured materials by inert gas condensation

EXERCISES

1. Consider spherical nanoparticles of copper with diameters of 10 and 100 nm.
   (a) Calculate the total number of atoms in the particles, considering the lattice parameter of bulk Cu.
   (b) Calculate the number of atoms at the grain boundaries for a typical grain boundary thickness of 1nm.
2. Define magic numbers.
3. Explain the reasons for change in the melting point of a nanoparticle with respect to its bulk melting point based on thermodynamic principles.
4. List commonly found nanomaterials in nature.
5. Write a brief note on nanophenomena in the biological world.
6. Give examples for 1D, 2D and 3D nanomaterials and describe their unique properties.
7. Explain the mechanism involved in self-cleaning textiles and windows.
8. Explain the mechanism of size dependence of colours in nanoclusters.
9. Write a short note on the negative impact of nanotechnology on society.
10. What is nano-bioinfo-cognizance?

Chapter 2

Unique Properties of Nanomaterials

Learning objectives

- Defects in nanocrystalline materials
- Effect of grain size on physical properties (melting point, elastic constants, diffusivity, magnetic, electrical, optical and thermal properties)
- Effect of grain size on mechanical properties (hardness, yield strength, ductility, toughness and creep)
- Grain growth behaviour in nanomaterials

As we approach nanoscale dimensions, we move closer to the atomic or molecular scales. Atoms are the building blocks of all matter. They can be assembled in many ways to obtain the desired product. Both the chemistry and the geometric arrangement of atoms can influence the properties of the material. Hence, if we have the ability to construct matter, atom by atom, we would be able to perform wonders. For example, we know that both graphite and diamond are made of pure carbon. Thus, in principle, if we are able to rearrange the atoms (carbon) in graphite at our discretion, it would be possible to make diamond! Or, if we could rearrange the atoms (silicon and oxygen) in sand (and add a few other trace elements), it should be possible to make a computer chip! Engineering at the nano-level can bring about large changes in the properties of the products. In Chapter 1, we saw how the high defect concentration in nanomaterials results in novel and unique physical, chemical, elastic and mechanical properties of this class of materials. A few of these are highlighted in this chapter.

2.1 MICROSTRUCTURE AND DEFECTS IN NANOCRYSTALLINE MATERIALS

In order to understand the novel properties of nanostructured materials, we need to understand the structure and its interrelationship with properties. The microstructural features of importance in nanomaterials include:

- Grain size, distribution and morphology
- The nature of grain boundaries and interphase interfaces
- Nature of intragrain defects
- Composition profiles across grains and interfaces
- Residual impurities from processing

Crystal lattice imperfections, such as point, linear, planar and volume defects, lead to the structure-sensitive properties of materials. The atomic structure of nanostructured materials

DEFECTS IN CRYSTALLINE MATERIALS

Crystals are three-dimensional, periodic arrangements of atoms/molecules in space. Any imperfection leading to disruption of periodicity is referred to as a 'crystalline defect'. These defects are usually classified based on their dimensionality, namely, point defects (0D), line defects (1D), surface defects (2D) and volume defects (3D). Vacancies and substitutional and interstitial solutes are the common point defects observed in metals and alloys. In case of ionic solids, Schottky (anion–cation vacancy pairs) and Frenkel (vacancy-interstitial pairs of the same ions) defects may also be observed. Dislocations are the most commonly observed line defects and refer to a missing plane of atoms. Among the surface defects, grain boundaries, twins, stacking faults and free surfaces are the most common. Inclusions, voids and microcracks constitute the volume defects.

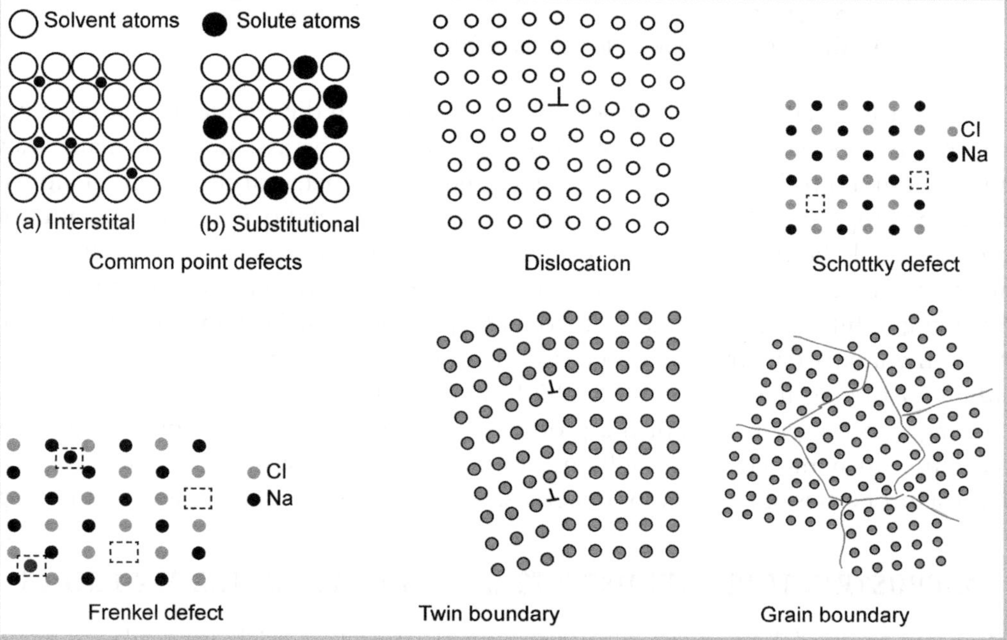

is unlike that seen in glass or conventional crystalline materials, because of the large volume fraction of grain boundaries and interfaces. Hence, in nanocrystalline materials, a substantial fraction of atoms lies at the grain boundaries and interfaces. It is not surprising that the behaviour of nanocrystalline materials is decided to a large extent by these defects, and as such, nanomaterials exhibit vastly different properties compared to bulk materials. The main defect types observed in nanocrystals are vacancies, grain boundaries, dislocations, twins, stacking faults and triple junctions.

2.1.1 Dislocations

Missing rows of atoms in a crystal are regions of high energy and stress due to disruption of the atomic bonds in the plane. This provides a driving force for dislocations to be annihilated at surfaces or grain boundaries to minimize the strain energy of the crystal. In effect, this may be treated as equivalent to an attractive force exerted by the surface on dislocations in the crystal. This force is inversely proportional to the distance of separation and hence becomes negligible for dislocations farther than a critical distance. However, for dislocations close to the surface or grain boundary, the attractive force can be large enough to result in annihilation of dislocations. Hence, for a small distance from the surface and grain boundaries, one would not expect to find any dislocations.

In order to treat this attractive force of surfaces and grain boundaries mathematically, a virtual image-dislocation of opposite sign is imagined to be existing at the surface. This problem can then be solved by considering the virtual force of attraction between the two dislocations of opposite sign. The force exerted by this virtual imaginary image dislocation on the dislocation defect in bulk is referred to as *image force* (F_{image}).

Dislocations are, in general, stable in conventional microcrystalline materials, though not thermodynamically stable defects. However, when the magnitude of the critical distance becomes comparable with that of the grain size, as in nanomaterials, the stability of dislocations is altered significantly. Hence, with decreasing grain size of nanograined materials, dislocation stability is reduced, due to the large grain boundary area. It is well known that dislocation mobility and interactions play a large role in determining the deformation and plastic flow behaviour of conventional crystalline materials. Hence, it is expected that the deformation behaviour of nanocrystalline materials is significantly different from that of conventional microcrystalline materials.

The typical critical size for dislocation stability in various metals is given in Table 2.1.

The typical dislocation density in annealed crystalline materials is about 10^{10}/cm. As the grain size is reduced to about 10 nm, the dislocation density can reduce by 2–3 orders or more and finally, below a critical grain size, dislocations are no longer stable, i.e., there will be no dislocations in the nanocrystalline materials below the critical grain size. However, in contrast to *whiskers* (single crystals without dislocations), nanomaterials have a large number of grain boundaries as defects. Hence, the plastic deformation behaviour of the material cannot be governed by dislocation mechanisms. This can result in significantly different mechanical

The condition for dislocation stability can be given by, $F_{image} \leq b\sigma_P$
where b is Burger's vector and σ_P is Peierl's stress.

The critical size (d_c) below which the image forces make the dislocations unstable is given by $d_c = Gb/\sigma_P$
where G is the shear modulus.

Table 2.1 Dislocation stability in typical BCC and FCC metals

(in nm)	Ni (FCC)	Fe (BCC)
dc (sphere)	16	3
dc (cylinder)	10	2

properties in nanomaterials below the critical sizes. The effect of decreasing dislocation density on the deformation mechanism is an area of significant scientific curiosity.

2.1.2 Twins, stacking faults and voids

Although thermodynamically metastable, planar defects are often observed, even after annealing, in many faceted nanomaterials, including nanorods and nanowires. These planar defects include twins and stacking faults (intrinsic or extrinsic), and are usually neglected by most analytical models. For example, many bulk metals have the face-centred cubic structure, but nanocrystals and nanorods of the same material often exhibit various structural modifications such as single or multiple symmetric twinning, as well as five-fold cyclic twinning, resulting in decahedral and truncated decahedral nanostructures below critical siges (see box). Figure 2.1 demonstrates the nano-twins in nanocrystalline Cu particles that are embedded in an amorphous Fe–Zr–B alloy. Twins are generally observed in crystals subjected to deformation under high strain rate or at low temperatures. During crystallization of liquid metal, it is expected that volume misfit strains can be easily accommodated in the liquid phase, and hence one does not anticipate the formation of twins in the nucleating crystals. Thus, the origin of such twins during the growth of nanoparticles from liquid is a subject of immense interest.

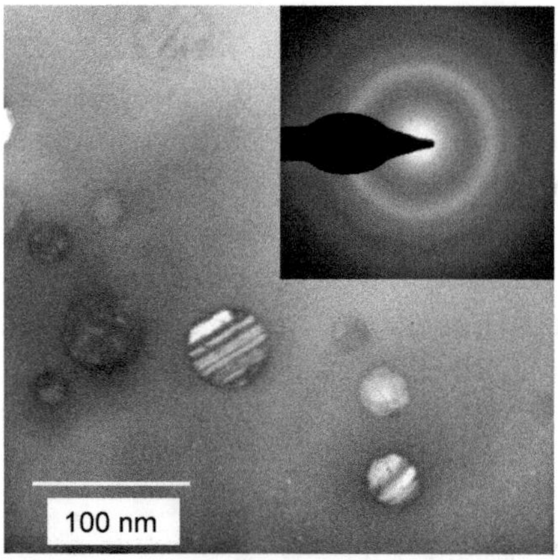

Fig. 2.1 Twins in nanocrystalline Cu particles embedded in amorphous Fe-Zr-B alloy. The twins formation is possibly due to strains developed at the amorphous Cu interface during solidification.
(*Source:* BS Murty, IIT Madras).

Voids in nanocrystallites may be situated at either triple junctions or as large porosities due to insufficient compaction and sintering of nanocrystallites synthesized from the powder route. Although both types of voids influence the behaviour of the nanocrystallites, the former is structurally more important. It is suggested that triple junction voids arise as a result of relaxation of nanocrystalline grain boundaries.

2.1.3 Grain boundaries, triple junctions and disclinations

A schematic representation of a hard-sphere model of an equiaxed nanocrystalline metal was shown in Chapter 1 (Fig. 1.9). Two types of atoms can be distinguished: crystal atoms with nearest-neighbour configurations corresponding to the lattice (black circles) and boundary atoms with a variety of

MULTIPLY TWINNED NANOPARTICLES

Multiply twinned particles (MTP) with a pseudo five-fold symmetry are observed frequently in nanocrystalline particles and thin films (deposited on crystalline substrates) of cubic face-centred metals, diamond-type semiconductors (C, Si, Ge) and alloys. MTP are formed by sequential cyclic five-fold twinning, with each twin sharing a common twin plane boundary with the adjacent twin. It is known, for example, that twinning in FCC occurs along the {111} plane. The angle between two {111} planes is 70.53 degrees. Thus, repeated twinning five times will still not result in 360 degrees, but will result in an annular gap of about 7.35 degrees. Thus, the five-fold symmetry formed from an FCC structure should have a distortion due to the geometrical arrangements along the five-fold axis. This results in a typical rotational defect, similar to disclination defects in such multiply twinned particles. The maximum size of these particles without distortion has been reported to be ~ 40 nm because the internal stresses of the crystals increase as they grow. The investigation of five-fold

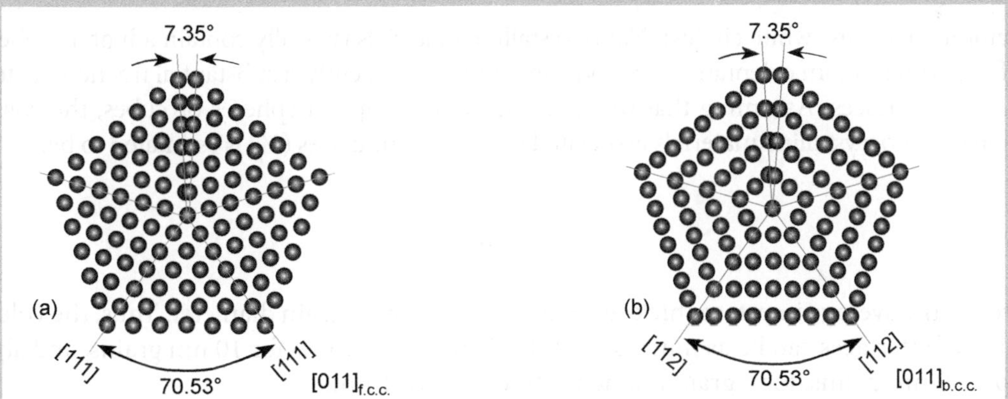

Schematic of five-fold twinned cubic lattice showing the annular mismatch created by repetitive twinning.

twinned structures in synthetic nanoparticles and thin films started in the second half of the 20th century. In 1957, Segall observed pentagonal grains of pyramidal shape in cold rolled Cu upon thermal etching. This was followed in 1959 by the observation of pentagonal whiskers (i.e., rod-like shapes) of Ni, Fe and Pt grown from the vapour phase on W substrates by Melmed and Hayward. Subsequently, MTPs have been observed in several metals, alloys and ceramic phases. Pseudo five-fold symmetry particles have been seen in microcrystalline chemical vapour deposited (CVD) diamond films and in various other nanoparticles of FCC, diamond cubic or BCC structure. MTPs of Ge and SnO_2 have been grown by physical vapour deposition processes. Nanocrystalline MTPs of c-BN, TiN, Si, SiC, etc., have been synthesized by CVD.

Physical and chemical properties of materials assembled from five-fold twinned nanoparticles may differ from materials consisting of untwinned nanoparticles in a variety of aspects, according to their respective structural characteristics. These differences concern properties sensitive to the surface energy, lattice symmetry, internal structure and surface structure, and they may cause changes in melting point, magnetic moment, electronic transition and chemical reactivity. For MTPs embedded in a matrix of foreign material, the interface structure has to be considered, which via particle–matrix interaction may influence the elastic properties of the composite. The appearance of spontaneous ferromagnetic order in Pd nanoparticles of about 6.8 nm size has been explained by a transition from single crystalline to multiply twinned structure with decreasing size.

interatomic spacings (white circles). Nanocrystalline materials typically contain a high number of interfaces with random orientation relationships, and consequently, a substantial fraction of atoms lies in the interfaces. Assuming that the grains have the shape of spheres or cubes, the volume fraction of nanocrystalline materials associated with the boundaries (V_i) is estimated to be:

$$V_i = \frac{3\delta}{d}$$

where δ is the average interface thickness and d is the average grain diameter. Thus, the volume fraction of interfaces can be as much as 60% for 5 nm grains, 30% for 10 nm grains, and about 3% for 100 nm grains, for a grain boundary thickness of 1 nm.

According to the phase mixture model, many properties of nanocrystalline materials can be estimated by a simple rule of mixtures,

$$X = V_{cr} X_{cr} + V_{ic} X_{ic}$$

where subscripts *cr* and *ic* refer to the crystalline and intercrystalline components of nanocrystalline materials, and X and V denote the property and volume fraction of the respective components. Intercrystalline components include the grain boundaries, triple lines and quadruple nodes. Taking a cubic unit cell or a regular polyhedron unit cell of nanocrystalline materials, the volume fraction of each component can be expressed as follows:

$$V_{cr} = \frac{(d - \delta^3)}{d^3}$$

$$V_{gb} = \frac{6\,(d-\delta)^2(\delta/2)}{d^3}$$

$$V_{ij} = \frac{12\,(d-\delta)(\delta/2)^2}{d^3}$$

$$V_{qn} = \frac{\delta^3}{d^3}$$

where the subscripts *cr, gb, tj* and *qn* refer to crystallite, grain boundary, triple lines and quadruple nodes, respectively; *d* and δ represent the grain size and grain boundary thickness.

Figure 1.10 in the previous chapter represents the variation of volume fraction of intercrystalline region with the grain size. When the grain size is smaller than about 20 nm, the total volume of the intercrystalline region (grain boundary and triple junctions) becomes significant. The density of grain boundaries in nanocrystals is very large (~ 10^{19} cm^{-3}) and there is wide distribution of interatomic spacing at these grain boundaries.

It has been suggested that the triple junctions can be described based on the disclination defect model. *Disclinations* (Fig. 2.2) are line defects characterised by a rotation vector ω in contrast to the translational vector *b* for dislocations. For example, consider a solid rubber ball. If you cut a slit in it of width only as thick as a blade and then try to forcefully bring together the ends of the cut portion, the resultant defect is a dislocation. Instead, if you cut a wedge in the ball and now try to bring the cut ends together, the resultant defect is a disclination. Unlike dislocations, the elastic stress fields of disclinations diverge with distance.

The triple junctions may be considered to form a network of disclinations. Theoretical calculations have shown that triple junction energies are comparable to dislocation energies, and that compensating disclinations play a significant role in the properties of nanocrystalline metals with grain size less than about 10 nm. These triple junctions are linear defects that play a significant role in the mechanical, thermodynamic and kinetic properties of polycrystals. This role may be particularly important in nanocrystalline materials, where the grain boundaries are short and contain a small number of structural units.

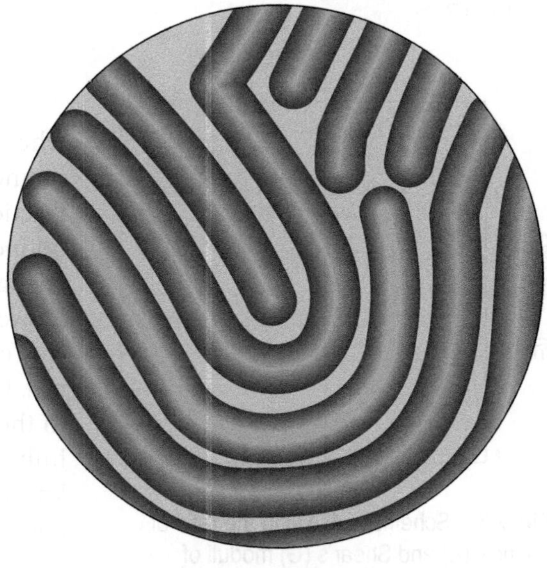

Fig. 2.2 A schematic showing disclination.

2.2 EFFECT OF NANO-DIMENSIONS ON MATERIALS BEHAVIOUR

2.2.1 Elastic properties

The elastic modulus of a material is proportional to the bond strength between atoms or molecules. The higher the bond strength, the higher will be the melting point and elastic modulus. The second differential of the interatomic force–distance curve at the position of equilibrium separation distance is known to be proportional to the elastic modulus. The elastic properties of crystalline materials are usually considered to be structure (microstructure) independent. If the temperature is increased, the mean separation between atoms increases and the elastic modulus decreases. A large increase in vacancy and other defect concentrations can be treated as equivalent to higher apparent temperature. Increasing the defect concentration is thus expected to decrease the elastic modulus. However, the effect of defects on the elastic modulus is manifested only at significantly higher concentration of vacancies. Nanomaterials, by virtue of their very high defect concentration, may have considerably lower elastic properties in comparison to bulk materials.

The first measurements of the elastic constants and, in particular, of Young's modulus, E, of nanocrystalline materials were performed with samples prepared by the inert gas condensation method. The elastic modulus of nanocrystalline compacts was found to be 30%–50% lower than bulk values (Fig. 2.3). One of the first explanations proposed was that the elastic moduli of the grain boundary regions are much smaller than the corresponding moduli of the grain bulk region, due to weaker bonds and lower density of atoms.

In terms of the linear theory of elasticity for polycrystalline materials, the elastic constants are expected to be reduced by a fraction determined by the volume fraction of the grain boundaries,

$$C^{-1} = (1 - \alpha)\, C_o^{-1} + \alpha\, C_{gr}^{-1}$$

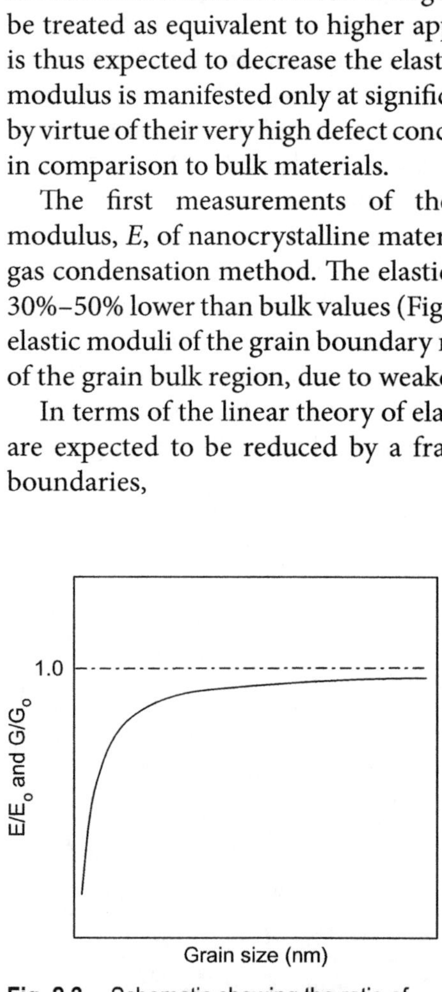

Fig. 2.3 Schematic showing the ratio of Young's (E) and Shear's (G) moduli of nanocrystalline materials to that of conventional grain size.

where C_o is the grain bulk modulus, C_{gr} is the grain boundary modulus; $\alpha \approx 3\Delta d/d$ is its relative volume fraction, where d is the grain size and Δd is the grain boundary width. This should lead to significant changes only for sizes about or below 10 nm. However, an anomalous elastic modulus reduction was observed also for greater sizes, up to d = 200 nm. However, it was soon realised that the porosities in the compact due to inadequate sintering can also influence the measured elastic properties. It is well known that elastic properties depend on the porosity content ($\Delta\rho_m$) according to the formula:

$$E = E_o \exp(-\beta\Delta\rho_m/\rho_m)$$

where β is a constant, while E and E_0 correspond to the measured and reference elastic modulus of the material, respectively. The effect of porosities on the elastic modulus is particularly predominant in materials with grain size below ~ 20 nm. The presence of porosities can result in a decrease in modulus of up to 20%–25%, and hence the effect of the intrinsic nature of nanocrystalline materials on the elastic modulus needs to be more carefully investigated.

By using a phase mixture model (consisting of a mixture of crystalline phase, intercrystalline phases, and pores), it was suggested that the elastic modulus of nanocrystalline materials should decrease with decreasing grain size. In contrast to nanograined materials, it has been suggested that the elastic modulus of carbon nanotubes increases with decrease in tube diameter. The increase of apparent elastic modulus for smaller diameters is attributed to the surface tension effects.

2.2.2 Melting point

As mentioned in Section 2.2.1, the decrease in bonding energy due to enhanced surface and grain boundary area in nanocrystalline materials can reduce the enthalpy of fusion and the melting temperature. At the melting point of a material, the solid and liquid phases are in equilibrium. For this to be thermodynamically possible, it is necessary that the chemical potential of a component (μ_i) is the same in both solid and liquid phases ($\mu_i^l = \mu_i^s$). However, due to inherent internal stresses, the pressure of the solid is greater than that of the liquid.

$$P_s = \frac{P_l + 2\gamma}{r}$$

where γ is the surface energy and r is the radius of the solid sphere. The chemical potential is related to pressure by the following equation:

$$d\mu = -SdT + VdP$$

By substituting the pressure relationship into the above equation and recalling that at equilibrium the chemical potential of the liquid is equal to that of the solid, assuming the overall pressure change in the liquid is negligible, and noting that $S_l - S_s = \Delta S_m$, the following equation is derived:

$$\Delta S_m dT = \frac{-2V\gamma\, dr}{r^2}$$

It should be recognised that the bulk melting temperature is related by:

$$\Delta S_m = \frac{\Delta H_m}{T_m}$$

Combining the above relations yields a relationship that describes the decrease in the melting point of a substance as the radius decreases.

$$\Delta T = \frac{2V\gamma T_m}{r\Delta H_m}$$

It can be seen that the change in melting temperature is inversely proportional to the radius of the sphere. In other words, as the grain size is reduced, the melting point is lowered.

It has been observed that nano-CdS of diameter ~ 2.5 nm melts at 600 K, much lower than the bulk melting point (1675 K). The single-walled carbon nanotube melts at ~ 1600 K, 0.42 times its bulk melting point (3800 K). A similar decrease in the melting point has also been made in case of nano-Si with a concurrent increase in hardness. At the surface of a nanosolid, the atomic coordination number (CN) is much lower than the standard atomic CN inside a bulk.

It is known that atoms in a solid vibrate about their mean position. The amplitude of the vibrations increases with increasing temperature. When the vibration amplitude exceeds a certain percentage of the bond length, melting begins at the surface and propagates through the solid. Atoms at the surface and grain boundary are less constrained to vibrate compared to atoms inside the crystal lattice. As the grain size decreases, the percentage of atoms residing at surfaces and grain boundaries increases significantly. Hence, free-standing nanoparticles may show a lower melting point compared to bulk. A similar effect has been reported on zinc nanowires embedded in holes in an anodic alumina membrane. The melting point of zinc nanowires was found to decrease with decreasing diameter of the nanowire.

In contrast to nanoclusters and nano-agglomerates, nanoparticles within a matrix may, in fact, experience an enhancement in the melting temperature. The matrix exerts a pressure, p, which can affect the melting temperature of the particles. This is described by:

$$p = \frac{2\mu\kappa\Delta V}{3V_o}$$

where μ is the shear modulus of the matrix, κ is a dimensionless factor that takes into account the presence of other particles, ΔV is the change in particle volume due to thermal expansion, and V_o is the initial particle volume. From the Clausius–Clapeyron equation ($dT/dP = \Delta V/\Delta S$, where T is the transition temperature, P is the pressure, ΔV and ΔS are the change in volume and entropy during transition), it is understood that the transition temperature will increase with increase in pressure, if change in volume on melting is positive. Hence, the melting point of nanoparticles embedded in a bulk matrix increases with decreasing size of particulate as the pressure increases with decrease in particle size (Fig. 2.4). Researchers at the University of California, Berkeley, USA, Lawrence Berkeley National Laboratory, USA and Australian National University, Australia, have found that Ge nanocrystals embedded in silica glass do not melt until temperatures are almost 200°C above the melting point of bulk

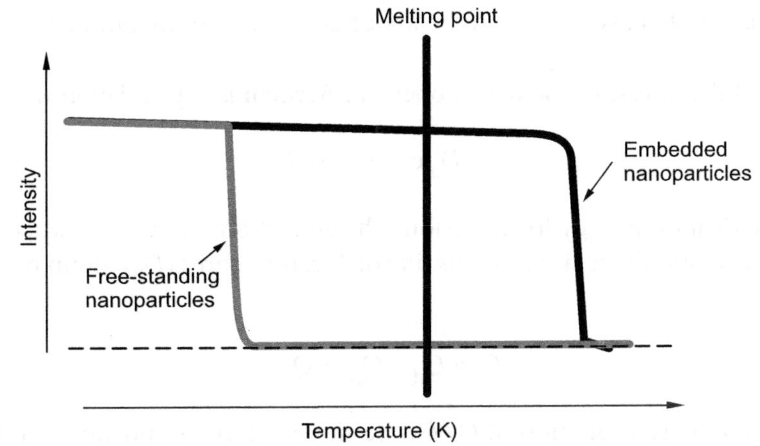

Fig. 2.4 Schematic diagram showing melting of free-standing nanoparticles and that of nanoparticles embedded in a higher melting matrix.

Ge, and resolidify only when the temperature is more than 200°C below its bulk melting point. Lead nanoparticles embedded in an Al matrix have also exhibited superheating.

It is fascinating to observe that the melting temperature does not continuously decrease with decreasing grain size in nano-dimensions. In fact, as the cluster size is reduced below a critical limit, the melting point of clusters is seen to increase above the bulk melting temperature of the material, at least in some cases. It has been found that a solid containing about 10 atoms of Ga or IV A elements (C, Si, Ge, Sn and Pb) melts at temperatures that are higher than the melting point of the corresponding bulk solid ($T_{m,b}$). This is in contrast to the behaviour of nanograined solids with grain size in the range of 1–100 nm, where a universal decrease in melting point is observed with the grain size. Compared to the melting point of bulk Ga, which is about 303 K, clusters of Ga_{39-40} melt at 550 K, while smaller clusters of Ga_{17} do not melt up to 700 K. Consistent insight into the phenomenon of melting point oscillation (suppression followed by elevation as the particle size is reduced from bulk to sub-nanometre size) over the whole range of sizes remains a scientific challenge.

2.2.3 Diffusivity

Diffusion kinetics increases with increasing defect content of the material, like vacancy concentration. Grain boundaries and dislocation cores provide easy diffusion paths compared to bulk lattice, because the structure is more open/defective. Thus, fewer atomic bonds have to be disrupted for diffusion along defect cores, resulting in lower activation energy paths. The numerous interfaces in nanocrystalline materials provide a high density of short circuit diffusion paths. Thus, nanocrystalline materials are expected to exhibit enhanced self-diffusivity in comparison to single crystals or conventional polycrystals with the same chemical composition. This idea was confirmed by self-diffusion measurements in nanocrystalline Cu. The measured diffusivities in nanocrystalline Cu are about 14–20 orders of magnitude higher than lattice diffusion. It is also known to be about 2–4 orders of magnitude larger than grain boundary self-diffusion.

It is known that the diffusion coefficient obeys an Arrhenius-type relation with temperature:

$$D = D_o \exp(-Q/RT)$$

where Q is the activation energy for diffusion. The activation energy for lattice diffusion (Q_l) is higher than that along short circuit paths like dislocation cores (Q_d), grain boundaries (Q_{gb}) and surfaces (Q_s).

$$Q_l > Q_d > Q_{gb} > Q_s$$

If log (D) is plotted as a function of ($1/T$), the slope will be proportional to the activation energy. Since the activation energy for grain boundary diffusion is much lower than that of lattice diffusion, the slopes of the plot are smaller for grain boundary diffusion (Fig. 2.5). For the same reason, the difference in the diffusion coefficients between the grain boundary and lattice are much higher at lower temperatures. It can be inferred from the figure that grain boundary diffusion will have a greater effect at lower temperatures. At higher temperatures, the diffusion coefficients of lattice and grain boundary diffusion are nearly the same and, considering the much lower volume fraction of grain boundaries in conventional materials, the effective diffusion coefficient would be largely determined by the lattice diffusion coefficient.

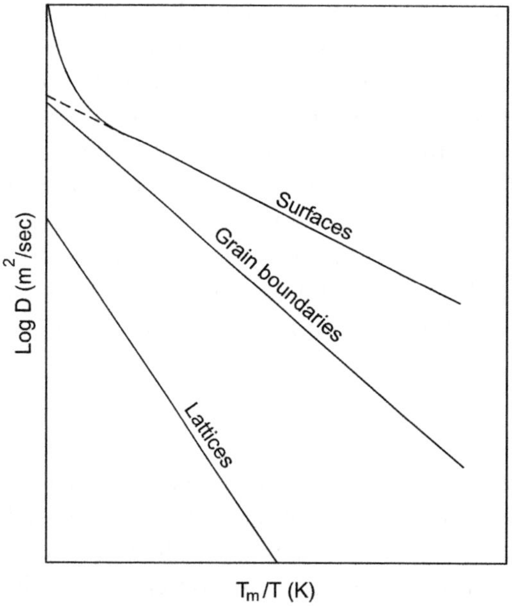

Fig. 2.5 Schematic showing diffusivity in metals along various diffusion paths. The higher diffusivity along the short circuit paths, such as grain boundaries and surfaces, is demonstrated in the figure.

Since the effective diffusion coefficient (D_{eff}) is an average of diffusion along different paths, it can be inferred that,

$$D_{eff} = D_l + (\delta/d)\, D_{gb} + f(V_d/d)\, D_p + f(V_s/d)\, D_s$$

where V_d and V_s are volume fractions of dislocation cores and surfaces in the material, respectively, d is the grain size, and D_l, D_p, D_{gb} and D_s are lattice, dislocation pipe, grain boundary and surface diffusion coefficients, respectively. δ is the grain boundary thickness and is usually about 1 nm. For a grain of about 1 μm in diameter, it is obvious that the second term in the above equation will be significantly less than D_l, meaning that the lattice diffusion coefficient will be nearly equal to the effective diffusion coefficient. However, as the grain sizes reduce to nano-dimensions, the second term has a dominating effect. Since grain boundary diffusion coefficients are a few orders higher than lattice diffusion, the effective diffusion coefficients of nanocrystalline materials are higher than that observed in coarse-grained materials. Table 2.2 demonstrates the enhanced diffusivities in nanomaterials.

Table 2.2 Diffusivities in nanomaterials (in m^2/s at 300 K)

System	Bulk	Glass	Nano
$^{67}Cu/Cu$	10^{-39}	—	10^{-19}
Ag/Cu	10^{-39}	10^{-36}	2×10^{-19}

2.2.4 Grain growth characteristics

Consider a sphere of radius r. Its surface area is proportional to the square of its radius, while the volume is proportional to the cube of the radius. Thus, the ratio of the surface area to the volume of a sphere, for example, can be seen to be inversely proportional to the radius. Thus, the smaller the crystallite size, the larger is the surface area per unit volume. Nanocrystalline materials have a significantly larger surface area per unit volume fraction of grain boundaries compared to microcrystalline solids.

Grain boundaries are higher energy regions in materials. Thus, there is always a driving force for reduction in the grain boundary surface area per unit volume. This can happen only by coarsening of smaller grains involving migration of grain boundaries. Such a reduction in grain boundary surface area per unit volume is the major driving force for grain coarsening in nanocrystalline materials, in order to reduce the net energy of the system. However, grain boundary mobility is predominantly a diffusion limited process and hence depends to a large extent on composition and temperature, amongst other considerations. Grain growth has an exponential dependence on the temperature as shown in the equation below:

$$r^n = r_0^{\,n} + \exp\left(\frac{-Q_{gb}}{RT}\right)$$

where r is the radius of the crystal, r_0 a material constant, Q_{gb} is the activation energy for grain boundary migration, and n is an exponent factor and is typically 2 for microcrystalline grains. Although the above expression has been derived for microcrystalline solids, it has also been found to be applicable to nanocrystalline materials. However, the activation energy for grain growth in nanocrystalline solids is expected to be lower. In addition, in nanocrystallline solids, owing to the higher driving force for grain growth (larger reduction in grain boundary surface area per unit volume possible by growth of nanocrystalline grains), the exponent factor, n, is normally much higher than 2. Both the above factors lead to enhanced grain coarsening kinetics in nanocrystalline solids.

Thus, it can be seen that nanocrystalline solids cannot be stable at high normalized temperatures (T/T_m) for any given material. If the nanocrystalline grains grow larger at high temperatures, all the advantageous properties of nanomaterials envisaged cannot be utilized in service. Thus, nanocrystalline solids are in general not suitable for high temperature applications. However, there is progress in grain boundary engineering to reduce grain coarsening kinetics of nanocrystalline materials. Grain boundary migration can be reduced by pinning the grain boundaries either with secondary particles (Zener pinning) or by vacancy clusters. Grain boundaries being higher energy sites, there is a driving force for several secondary phases to be located there. Grain boundary migration would thus incur an additional energy to break the bonding with such particles, acting as a source of inhibition for grain growth. The effective pinning ability of secondary phases is obviously dependent on both the volume fraction as well as the size of the particle.

It is clear that for effective Zener pinning, the particles should be stable at high temperatures and should not undergo coarsening themselves. This can also be achieved by having dispersoids of secondary phases like ceramics (that are insoluble in the matrix even at elevated temperatures) decorating the grain boundaries of nanocrystalline materials. An effective means of synthesizing such a solid is by powder metallurgy processes involving ball milling and consolidation techniques. If the dispersoid phase is insoluble in the matrix, it is obvious that the ceramic second phase would be both stable at high temperatures and will not undergo coarsening of itself. Thus, such a technique can be an effective route for achieving grain boundary pinning to enhance the service temperature applicability of nanomaterials. Figure 2.6 (see Plate 4) demonstrates the Zener pinning effect of W on grain growth in Cu–W nanocomposites. Copper nanograins do not significantly grow in the presence of tungsten nanoparticles, in spite of heating it to 0.6 of its melting point.

There is another way of grain boundary engineering to restrict grain growth. This involves enabling a grain boundary with a composition different from that of the bulk. Thus, grain boundary migration will also have to simultaneously involve diffusion of all involved chemical species to ensure the same composition of grain boundary. Such a solute drag effect can act as a rate limiting step for grain boundary migration. A heterogeneous composition can be obtained, for example, by grain boundary segregation effects. The solute drag effect of Cu in Ag in controlling the grain coarsening tendency of nano-silver is demonstrated in Fig. 2.7. In this case, nanoparticles of pure Ag obtained by inert gas condensation of Ag clusters on a substrate grow significantly when heated to 600°C. In contrast, when Cu (having a higher melting point than Ag) is alloyed with Ag, the grain growth is significantly reduced at the same temperature due to solute drag effect.

2.2.5 Enhanced solid solubility

The solubility of a solute A in a solvent B is controlled by the chemical potential, μ_A, of A in B. The chemical potential and hence the solubility of A in B may be enhanced (or reduced) in nanocrystalline materials in comparison to single crystals or glasses with the same chemical composition. The solubility of H in nanocrystalline Pd (at concentrations $<10^{-3}$) is increased by a factor of 10 to 100 relative to a Pd single crystal. Similar effects have been observed in Mg and with the Mg-based alloys that have become popular for hydrogen storage applications. Figure 2.8 shows a schematic of how the hydrogen adsorption increases with a decrease in particle size. This result has significant technological importance.

Fig. 2.7 Grain growth restriction in Ag–Cu nanoparticles: (a) Ag nanoparticles heated to 600°C and (b) Ag–Cu nanoparticles heated to 600°C. The figure clearly shows that alloying Ag with the higher melting Cu reduces the diffusivity, which leads to grain growth inhibition due to solute drag effect in single-phase alloys. This is different from the Zener pinning effects observed in two-phase alloys. (*Source:* BS Murty, IIT Madras).

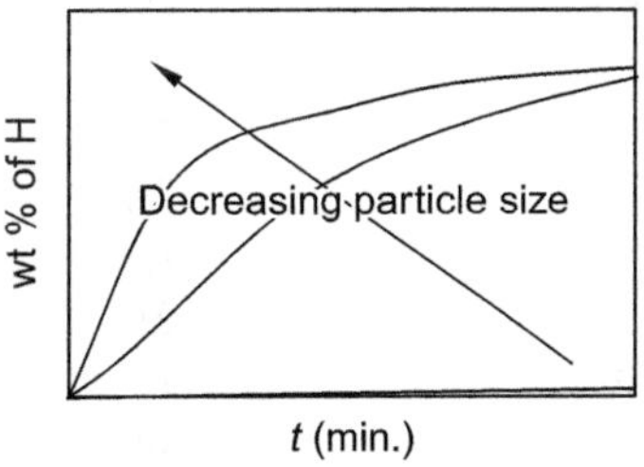

Fig. 2.8 Schematic showing hydrogen adsorption in nanocrystalline material with decreasing grain size.

A similar effect was observed for Bi in Cu. The solubility of Bi in crystalline Cu is less than 10^{-4} at 100°C. In contrast, the solubility of Bi in nanocrystalline Cu is about 4%, corresponding to a solubility enhancement by about 10,000 times relative to crystalline Cu. It is found that a diameter of 20 nm seems to be the threshold value to observe significant extended solid solubility, i.e., the solubility being greatly promoted with decreasing grain size when the size of the matrix nanoparticles is less than 20 nm. Figure 2.9 demonstrates this phenomenon in the Fe–Si system. It can be clearly seen that the solubility of Si in Fe is almost negligible up to a crystallite size of Fe of about 20 nm, below which there is a sudden dissolution of Si. The figure also shows that this solubility can be correlated to the grain boundary volume in nanocrystalline materials, which follows a similar trend. It has also been possible to alloy conventionally immiscible systems by optimising the process conditions to achieve nanocrystalline grains. Table 2.3 gives the extension of terminal solid solubility achieved on nanocrystallization by mechanical alloying (high-energy ball milling technique).

2.2.6 Magnetic properties

The study of magnetic nanoparticles has seen a large research effort in recent years. The main driving force for this increased interest is their application potential in vital areas like ultrahigh density magnetic storage devices, magnetic random access memory (MRAM), ferrofluids, spintronics, magnetic semiconductors, nanogranular magnetic materials, etc. With this increased interest comes the realization that the shape and size of the nanoparticles

Table 2.3 Extension of terminal solid solubility on nanocrystallization by mechanical alloying (MA)

Solvent	Solute	Solid solubility at room temperature (at.%) Equilibrium	MA
Ag	Cu	0.3	100.0
	Ni	0.7	3.8
Al	Cr	0.0	5.0
	Fe	0.0	4.5
	Mg	2.1	23.0
	Mn	0.0	18.5
	Nb	0.0	25–30
	Ti	0.0	6.0
	Zr	0.0	9.1
Cd	Zn	0.0	50.0
Co	C	0.0	6.0
	Cr	0.0	40.0
	V	9.1	33.0
	Zr	0.0	5.0
Cr	Co	9.0	40.0
	Cu	0.0	20.0
Cu	Co	0.0	90.0
	Fe	0.0	50.0
	Hg	0.0	70.0
	Zn	30.0	50.0
Fe	Al	18.5	50.0
	Cu	0.0	15.0
	Mg	0.0	20.0
	Si	9.0	27.5
Mg	Ti	0.0	4.2
Mn	Co	4.0	50.0
Nb	Al	21.5	60.0
	Cu	0.0	20.0
	Ni	7.0	10.0
Ni	Ag	2.0	9.0
	Al	10.0	27.0
	C	0.0	12.0
	Nb	6.0	15.0
Ti	Al	36.0	55.0
	Cu	0.0	8.0
	Mg	2.9	60.0
V	Co	7.0	40.0
Zr	Al	0.5	15.0
	Co	0.0	4.0

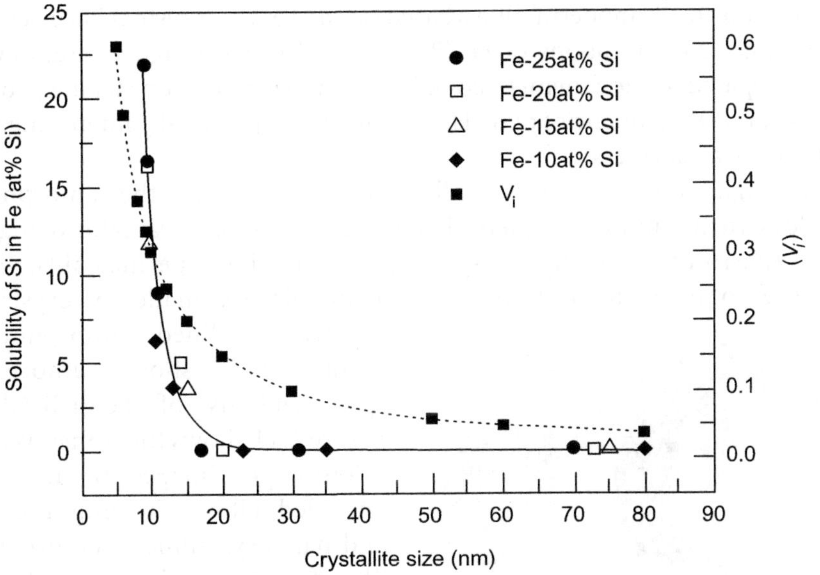

Fig. 2.9 Solubility of Si in Fe as a function of crystallite size of Fe. The figure demonstrates significantly higher solubility of Si in nanocrystalline Fe below 20 nm. (*Source:* BS Murty, IIT Madras).

take centre stage in determining their magnetic properties. The strength of a magnet is decided by its coercivity and saturation magnetization values. They increase when the grain size decreases and the specific surface area per unit volume of the grains increases.

Soft magnetic nanocrystalline alloys

In conventional soft magnetic alloys, low coercivity has been achieved using coarse-grained materials so that magnetic flux pinning at the grain boundaries is avoided. In nanocrystalline materials, when the grain sizes are much smaller than the domain wall width, the magnetic anisotropy is averaged over many grains and orientations and hence the coercivity is significantly reduced and permeability is enhanced.

Fe–Si–B–Nb–Cu amorphous alloys are found to transform to a BCC Fe–Si solid solution with grain sizes of about 10 nm during annealing at temperatures above the crystallization temperature. The presence of small amounts of Cu helps to increase the nucleation rate of the BCC phase, while Nb retards grain growth. These 'Finemet' alloys provide low core losses (even lower than amorphous soft magnetic alloys such as Co–Fe–Si-B), exhibit saturation induction of about 1.2 T, and exhibit very good properties at high frequencies, comparable to the best Co-based amorphous alloys. Though the Finemet-type nanocrystalline alloys are superior in their magnetic properties, they exhibit lower saturation induction than Fe-metalloid amorphous alloys, mainly because of the lower Fe content to attain amorphization and because of the addition of Nb and Cu (or other elements to control the nucleation and growth kinetics).

To avoid this problem, 'Nanoperm' alloys based on the Fe–Zr–B system have been developed. These contain larger concentrations of Fe (83%–89%) and have higher saturation induction (~1.6–1.7 T) compared to the Finemet alloys. The Nanoperm nanocrystalline alloys have very low energy loss at power frequencies of 60 Hz, making them potentially interesting for electrical power distribution transformers.

The small single-domain nanocrystalline Fe particles in the amorphous matrix give these alloys their unique magnetic behaviour, the most dramatic being the lowest energy losses (narrowest B/H hysteresis loop) of any high permeability. These alloys exhibit nearly zero magnetostriction. These materials are made by crystallization of rapidly solidified amorphous ribbons Figure 2.10 shows amorphous melt spun ribbons of Fe-Si-B-Nb-Cu alloy. Figure 2.11 shows the transmission electron microscopic images of melt spun and annealed alloys showing the amorphous and nanocrystalline structure, respectively. The decrease in hysteresis loop size due to nanocrystalline phase formation on annealing is shown in Fig. 2.12.

Fig. 2.10 Amorphous ribbon of Fe–Si–B–Nb–Cu alloy. (*Courtesy:* Bhaskar Majumdar, DMRL, Hyderabad).

Permanent Magnetic Nanocrystalline Materials

The nanoscale two-phase mixtures of a hard magnetic phase and a soft magnetic phase

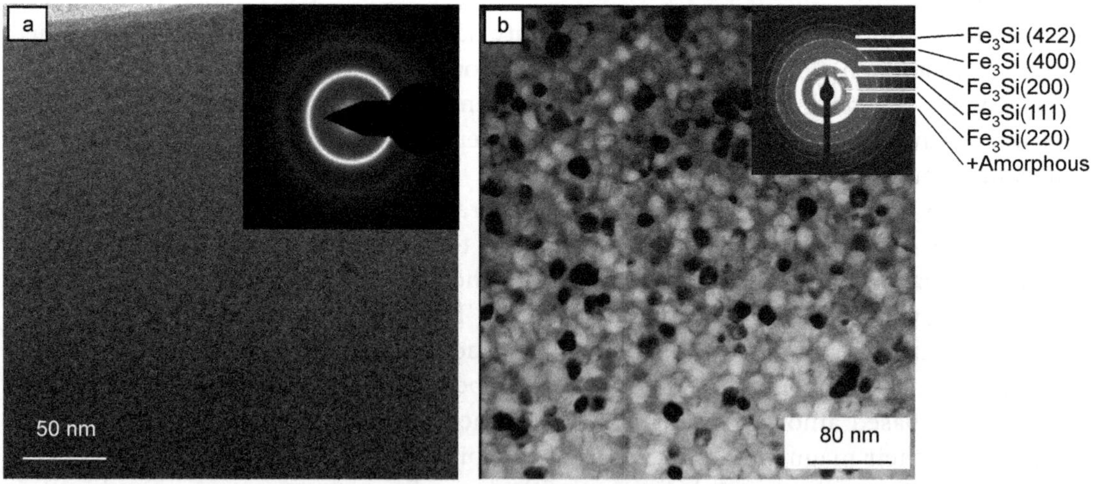

Fig. 2.11 Bright field transmission electron microscopy (TEM) images and corresponding selected area diffraction (SAD) patterns of (a) as melt spun and (b) annealed (525°C for 1 h) ribbons of Fe–Si–B–Nb–Cu alloy. (*Courtesy:* Bhaskar Majumdar, DMRL, Hyderabad).

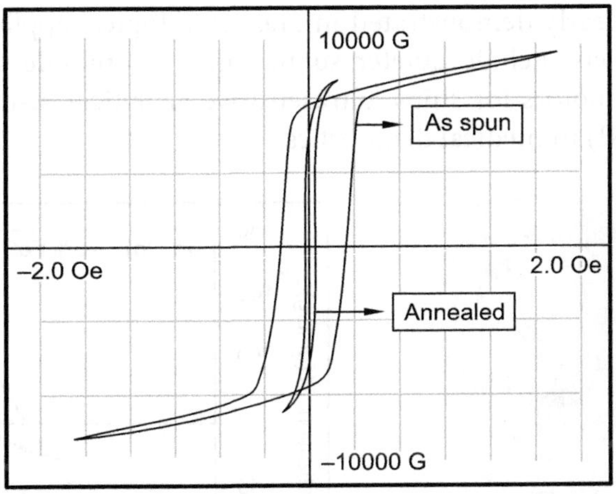

Fig. 2.12 Hysteresis loops of the as-spun and annealed Fe–Si–B–Nb–Cu ribbons. (*Courtesy*: Bhaskar Majumdar, DMRL, Hyderabad).

can exhibit values of remanent magnetization, M_r, significantly greater than the isotropic value of $0.5M_s$. Enhancement of remanence in such a composite can be obtained when the nanocrystalline grain size and the degree of coherence across interphase boundaries are such that exchange coupling occurs between the two phases. Exchange coupling allows crystallographically isotropic materials to exhibit remanence values approaching those achieved after full alignment (Fig. 2.13). The Fe-rich compositions (e.g., $Fe_{90}Nd_7B_3$) result in a mixture of the hard $Fe_{14}Nd_2B$-phase and a soft Fe-phase, which increases the magnetic induction. These two-phase nanoscale ferromagnetic alloys have been prepared by non-equilibrium methods such as melt spinning, mechanical alloying and sputter deposition. Magnets made of nanocrystalline Y–Sm–Co grains possess unusual magnetic properties due to their extremely large surface area. The magnetic properties of the Sm_2Co_{17} hard magnet can be significantly improved by developing nanocomposites with either Fe or FeCo alloy

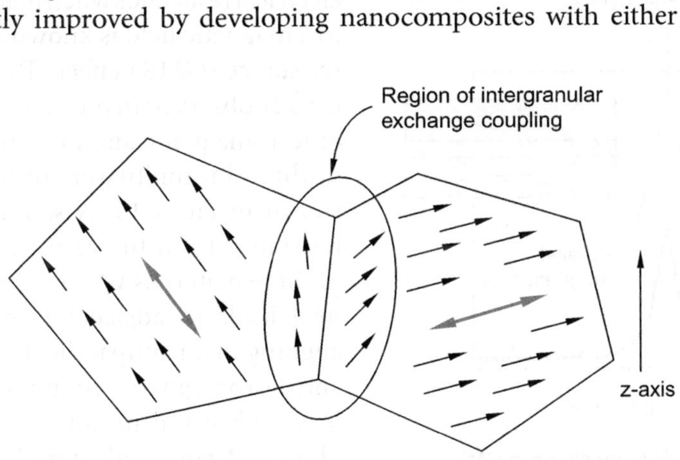

Fig. 2.13 Schematic diagram explaining the principle of exchange coupling in magnetic nanocomposites.

soft magnet, which is clearly demonstrated in Fig. 2.14. Typical applications for these high-power rare earth magnets include quieter submarines, automobile alternators, land-based power generators, and motors for ships, ultrasensitive analytical instruments and magnetic resonance imaging (MRI) in medical diagnostics.

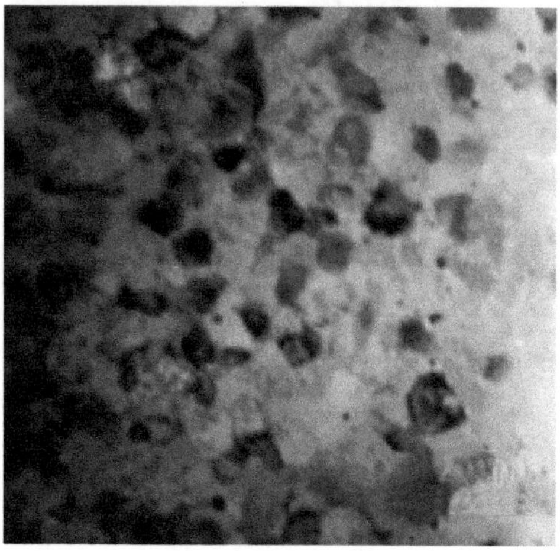

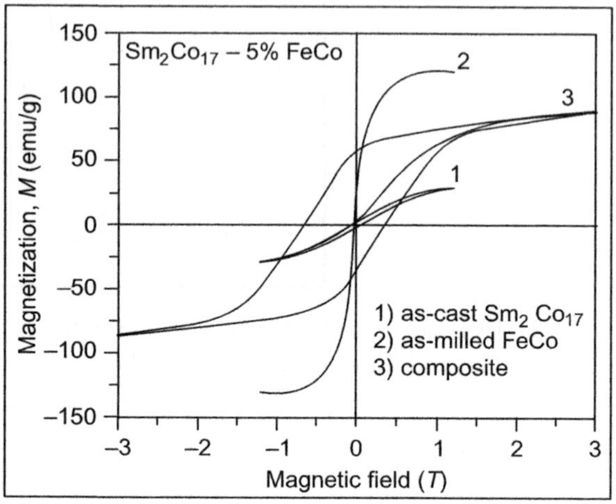

Fig. 2.14 Improvement in magnetic properties in Sm_2Co_{17}–FeCo nanocomposites in relation to bulk microcrystalline composites. The improved magnetic properties can be attributed to exchange coupling between the soft and hard magnetic phases in the nanocomposites. (*Source:* BS Murty, IIT Madras).

Giant magnetoresistance (GMR)

The phenomenon of significant decrease in electrical resistance when materials are exposed to a magnetic field is known as giant magneto-resistance (GMR) effect. This phenomenon is usually observed in bulk composites consisting of ferromagnetic and non-magnetic phases or in thin film multilayers of these materials, as shown in Fig. 2.15. A significant decrease in resistance from the zero-field state is evident in these materials when the external magnetic field leads to adjacent ferromagnetic layers, aligning in an antiparallel fashion due to weak anti-ferromagnetic coupling between layers. It was observed in both nanocrystalline thin films and equiaxed granular nanocrystalline materials.

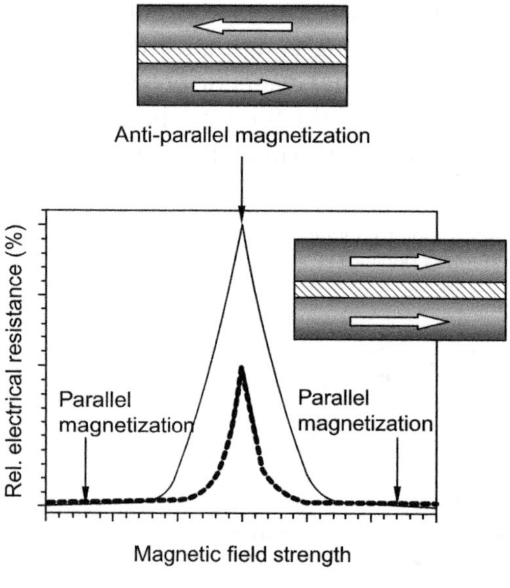

Fig. 2.15 Schematic diagram showing GMR effect in magnetic ultrathin films.

In case of granular materials, GMR is observed when small ferromagnetic single-domain particles with randomly oriented magnetic axes are embedded in a non-magnetic matrix. The explanation for the GMR is spin-dependent scattering of the conduction electrons at the ferromagnetic/non-magnetic interfaces and, to a lesser extent, within the magnetic grains. The GMR is inversely proportional to the average particle diameter. The effect can be up to 100% in multiple stacks of ultrathin films. This effect was first observed in Fe/Cr multilayer ultrathin films and is also commonly observed in Cu/Co nanocomposites. This effect has wide application in magnetic reading heads for computer hard discs and in position sensors.

2.2.7 Electrical properties

Nanomaterials can hold considerably more energy than conventional coarse-grained materials because of their large grain boundary (surface) area. They are materials in which an optical absorption band can be introduced, or an existing band can be altered by the passage of current through these materials, or by the application of an electric field. Conventional and rechargeable batteries are used in many applications that need electrical energy. The energy density (storage capacity) of these batteries is usually quite low, requiring frequent recharging. Nanocrystalline materials are good candidates for separator plates in batteries because they can hold considerably more energy than conventional ones. Nickel–metal hydride batteries made of nanocrystalline nickel and metal hydrides are envisioned to require far less frequent recharging and to last much longer. The dielectric properties are significantly enhanced by making these nanocrystalline. The dielectric constant of lead zirconium titanate (PZT) can be increased to 35,000 when it is made in the nanocrystalline state in comparison to a value of 2000 in the microcrystalline PZT (Fig. 2.16). When ferroelectric and ferromagnetic phases are brought together in a nanocrystalline state, there

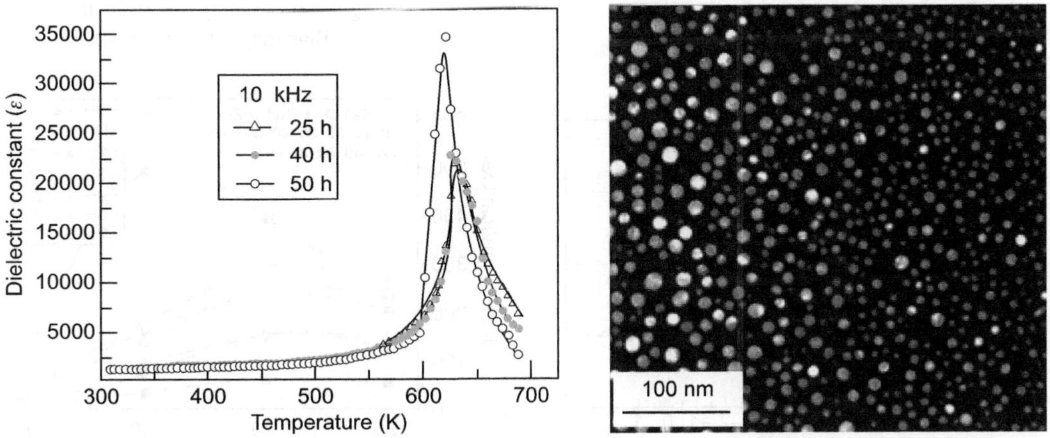

Fig. 2.16 Exceptionally high dielectric constant in nanocrystalline PZT. The nanocrystalline PZT with an average particle size of about 10 nm shows a dielectric constant of about 35,000 at T_c. In contrast, bulk PZT has a dielectric constant of only about 1000 at the same temperature. (*Source:* BS Murty, IIT Madras).

can be high magneto-electric effect in the material, which is demonstrated in Fig. 2.17. There are examples where the electrical nature of materials changes on nanocrystallization. Figure 2.18 is an example for this behaviour; a metallic Al–Cu–Fe alloy is converted to a semiconductor with a negative temperature coefficient of electrical resistivity in the nanoquasicrystalline phase, the explanation for which is still inconclusive, providing scope for further work.

2.2.8 Optical properties

Nanocrystalline systems have attracted much interest due to their novel optical properties, which differ remarkably from bulk crystals. Key contributory factors include quantum confinement of electrical carriers within nanoparticles, efficient energy and charge transfer over nanoscale distances and, in many systems, a highly enhanced role of interfaces. The linear and non-linear optical properties of such materials can be finely tailored by controlling the crystal dimensions, and the chemistry of their surfaces. Fabrication technology becomes a key factor for the applications.

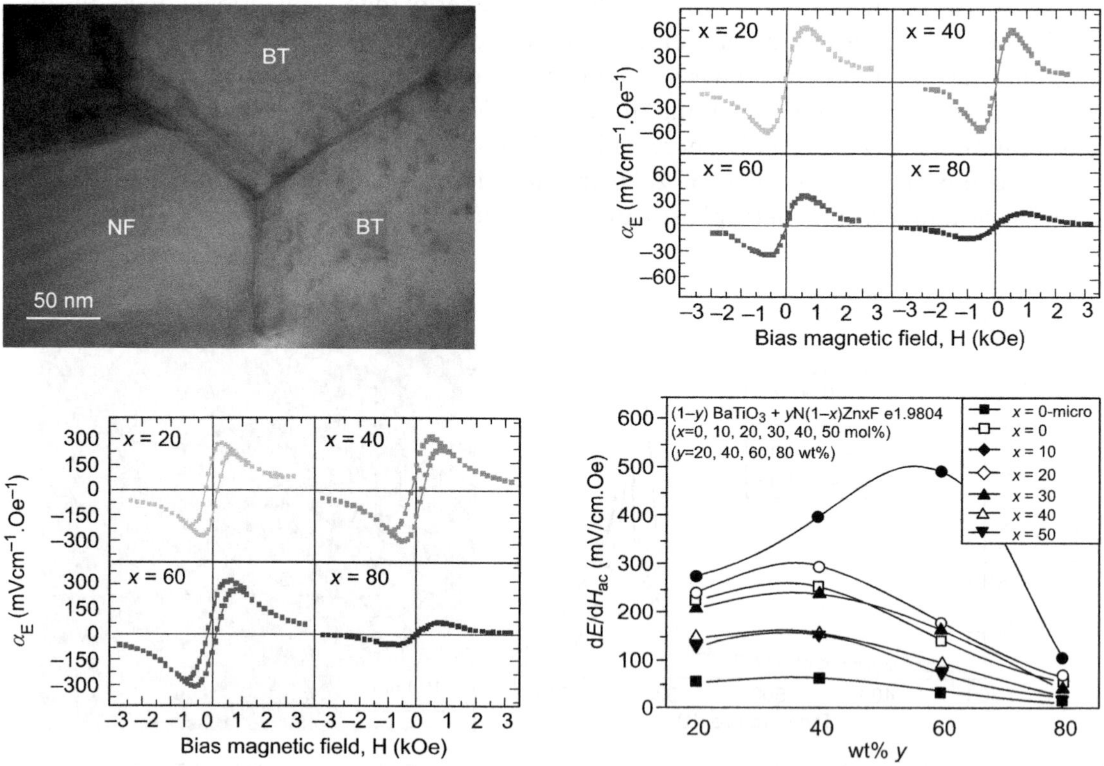

Fig. 2.17 The high magneto-electric effect in nickel ferrite/nickel zinc ferrite–barium titanate nanocomposite. (*Source:* BS Murty, IIT Madras).

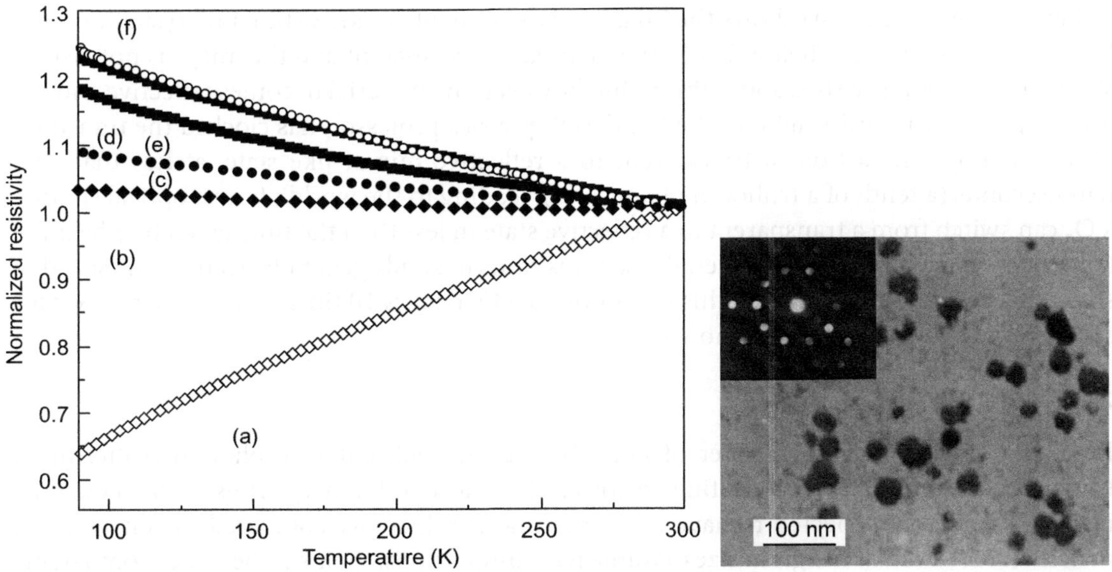

Fig. 2.18 Al–Cu–Fe alloy being converted from metallic to semiconducting nature on nano-quasicrystallization. (a) Bulk Al–Cu–Fe alloy, (b)–(f): Different fractions of nano-quasicrystalline phase (*Source:* BS Murty, IIT Madras).

Plasmons These are quasiparticles obtained by quantization of plasma oscillations, similar to *photons*, which form by quantization of light and sound waves. Plasmons can couple with a photon to create another quasiparticle called *plasma polariton. Surface plasmons* (SP) are plasmons that are confined to surfaces and interact strongly with light resulting in a polariton. SPs are responsible for the colour of nanomaterials. An SP is a natural oscillation of the electron gas inside a given nanosphere. If the nanosphere is smaller than the wavelength of light, the frequency of which is close to that of the SP, then the SP will absorb energy. The frequency of the SP depends on the dielectric function of the nanomaterial, and the shape of the nanoparticles. For spherical particles of gold, the frequency is about 0.58 of the bulk plasma frequency. Thus, although the bulk plasma frequency is in the UV region, the SP frequency is in the visible range (wavelength close to 520 nm). Suppose we have a suspension of nanoparticles in a host and a wave of light is applied, the local electric field may be hugely enhanced near an SP resonance. If so, one expects various non-linear susceptibilities, which depend on higher powers of the electric field to be enhanced even more.

The ultrafast optical emission of nano-sized diamond crystallites has been detected under the picosecond (ps) laser excitation of 300 nm through ultrafast fluorescence spectroscopy. This optical emission and photoluminescence (PL), induced at a laser wavelength of 300 nm, confines blue light. It has not been observed under the same experimental conditions from samples of a natural single crystal. The PL phenomenon of nano-diamond can be understood considering the effects of the surface states and the enlarged surface-to-volume ratios in nanoscale materials.

Figure 2.19 (see Plate 4) shows the image of Don Quixote made with nanocrystalline VO_2. When the temperature is below 341 K, the material is transparent and the image is not visible. When the temperature rises above this point, however, the material becomes reflective and the image appears. Richard Haglund, the Vanderbilt physics professor, has clocked the transition of VO_2 nanoparticles from a transparent to a reflective mirror-like state, at less than 100 femtoseconds (a tenth of a trillionth of a second), leading to the world's fastest optical shutter. VO_2 can switch from a transparent to a reflective state in less than the time taken by a beam of light to travel a tenth of a millimeter. Phase transitions in solids generally occur at the speed of sound in the materials; however, this transition in VO_2 occurs 10 times faster. The reasons for such a rapid change are not yet known.

2.2.9 Thermal properties

In general, increasing the number of grain boundaries will enhance phonon scattering at the disordered boundaries, resulting in lower thermal conductivity. Thus, nanocrystalline materials would be expected to have lower thermal conductivity compared to conventional materials. However, as the grain sizes assume nanodimensions, their size becomes comparable to the mean free paths of phonons that transport thermal energy. Thus, nanomaterials can show widely different properties compared to coarse-grained materials, due to the photon confinement and quantization effects of photon transport. It has been observed that in addition to the grain size, the shape also has an influence on the thermal properties of nanomaterials. For example, one-dimensional nanowires may offer ultralow thermal conductivities, quite different from that of carbon nanotubes. In nanowires, quantum confinement of phonons in 1D can result in additional polarization modes compared to that observed in bulk solids. The strong phonon–phonon interactions and enhanced scattering at grain boundaries result in a significant reduction in thermal conductivity of nanostructures. Silicon nanowires are known to exhibit thermal conductivity at least about two orders of magnitude smaller than that of bulk silicon. In contrast, the tubular structures of carbon nanotubes result in an extremely high (~ 6600 W/mK) thermal conductivity along the axial direction. However, high anisotropy in their heat transport property is observed, making the thermal transport direction-dependent.

In multilayered coatings, many collective modes of phonon transport may appear besides the phonon modes in each single layer; when the phonon coherence length becomes comparable to the thickness of each layer, the transport properties are significantly influenced. When the mean free path of phonons spans multiple interfaces, the phonon dispersion relation is modified, resulting in enhanced scattering due to decrease in phonon group velocity. Further, if the multilayer is designed to have a superlattice structure, and alternate films have a large mismatch in the phonon dispersion relations, it is possible that phonons in a certain frequency range may not propagate to the neighbouring layers unless there are mode conversions at the interface. Also, the presence of interface dislocations and defects can contribute to enhanced boundary scattering. All these factors can contribute to the lower thermal conductivity of multilayered nanostructured films.

The use of a nanofluid to enhance thermal transport is another promising application of the thermal properties of nanomaterials. Nanofluids represent the class of liquids that have a stable colloidal dispersion of nanoparticles distributed uniformly in the medium. It has been observed that dispersion of a wide variety of nanoparticles of oxides, nitrides, metals, metal carbides and nanofibres, such as single- and multi-walled carbon nanotubes, can significantly enhance the thermal conductivity of the fluid. To obtain stable colloidal suspensions, the particle size should normally be in the range of 1–100 nm and an anti-coagulant may also be added to enhance the stability of the nanofluid. The idea of enhancing the thermal conductivity of liquids using solid dispersions is not completely new. Maxwell proposed the possibility of enhancing the thermal conductivity of fluids using particle dispersions as early as 1873. This is because we know that solids in general have much higher thermal conductivity than liquids and gases. For example, while air has a conductivity of the order of 0.03 W/mK, water has a conductivity of the order of 0.6 W/mK, and metals like silver and copper have a conductivity of 400 W/mK, while materials like carbon nanotubes (CNT) have extremely high conductivity of 2300 W/mK.

Thus, it is obvious that thermal conductivity of a fluid can be enhanced by having particles of better heat transport properties dispersed in it. However, in the early years of development of particle dispersed fluids, microcrystalline dispersoid particles were used. These fluids suffered from inferior stability of suspension, leading to their coagulation and precipitation. Erosion of the walls of the pipes by the particles was also observed to be a major problem. With the advent of techniques to synthesize nanoparticles with controlled grain size, nanofluids with improved stability have been developed.

For dilute fluids with spherical dispersoids, the effective thermal conductivity (k_{eff}) of the fluid is expected to follow the equation:

$$\frac{k_{eff}}{k_m} = 1 + \frac{3\phi}{[(k_p + 2k_m)/(k_p - k_m) - \phi]}.$$

where k_p and k_m are the thermal conductivity of the particles and the medium, respectively, and ϕ is the volume fraction of the particles. The relation suggests that the thermal conductivity of a fluid will increase by about three times the effect expected from volume fraction effects. However, it may be noted that this equation does not suggest the dependence of thermal conductivity of the fluid on the particle size. This relation does not also predict particle-specific properties of the fluid. However, in practice, the thermal properties of nanofluids are known to depend on particle size, volume fraction and the type of dispersoid employed.

The first heat transfer enhancement with nano-sized particles was reported by Masuda and his group in Japan and they demonstrated that the thermal conductivity of ultrafine suspensions of alumina, silica and other oxides in water increased by a substantial amount (maximum of 30%) for a particle volume fraction of 4.3%. Choi and his group at the Argonne National Laboratory proposed to construct a new class of engineered fluids with superior heat transfer capabilities in 1995. Since then, a series of experiments have been performed with nanofluids. Incidentally, the term 'nanofluid' was first coined by Choi for denoting this new class of engineered fluids. An approximately 20% improvement in effective thermal conductivity has been reported when 5 vol.% CuO nanoparticles are added to water. The effective thermal conductivity of ethylene glycol increased by 40% for a nanofluid containing approximately 0.3 vol.% Cu nanoparticles of mean diameter <10 nm. High thermal conductivities in nanofluids can be due to a number of factors such as Brownian motion of the nanoparticles, molecular-level layering of the liquid at the liquid/particle interface, the nature of heat transport in the nanoparticles and the effects of clustering of nanoparticles. It is postulated that ballistic, rather than diffusive mechanisms account for the rapid heat transport in nanofluids.

Apart from enhancing the thermal conductivity of the medium, the use of nanofluids can also aid reduction in pumping power to attain equivalent heat transfer coefficient. An increase in pumping power by a factor of 10 results in doubling of thermal conductivity of conventional fluids. In contrast, nanofluids can achieve improved thermal transport properties

BALLISTIC AND DIFFUSIVE HEAT TRANSPORT

Heat transfer, in general, occurs by vibrations of atoms in solids. Based on quantum mechanics, the vibrations in solids can be mathematically treated as a convolution of phonons (normal modes) and photons (normal vibrations). Generally, heat transfer occurs by diffusive mechanism, which follows Fourier's equation for conduction and convection. However, when the particle size becomes comparable to the mean free path of phonons, ballistic heat transfer mechanism becomes predominant. For example, the mean free path of a phonon in a Cu particle ranges from 20 nm at 323 K to 16 nm at 423 K. When the particle size becomes smaller than these values, ballistic heat transport mechanism takes over. Under these circumstances, the heat flux would be proportional to the fourth power of temperature rather than linear dependence observed in diffusive heat transfer. This is depicted in the table given below, where q is heat flux, k is thermal conductivity, l is the mean free path of a phonon and L is the grain/particle size.

Transport type	Size scale	Governing equation	Schematic diagram
Diffusive	$L >> l$	$q = k\nabla T$	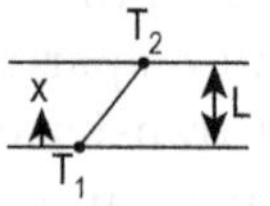
Ballistic	$L = l$	$q = \sigma (T_1^4 - T_2^4)$	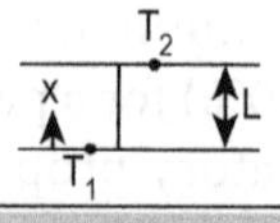

without enhancing the pumping power. Therefore, the potential savings in pumping power is significant with nanofluids.

2.2.10 Mechanical properties

The mechanical behaviour of nanocrystalline materials is being widely researched, because of their radically different properties. The tensile, work hardening, creep, fatigue and deformation behaviour of nanocrystalline materials are significantly different from that of their bulk counterparts due to the complex interaction of the defect structures. The major factors contributing to the difference in the mechanical behaviour of nanostructured materials are the following:

1. Very high fraction of atoms residing at grain boundaries and triple junctions exert a strong influence on the behaviour of these defects on application of stress. The influence of grain size on dislocation stability, as discussed in an earlier section, also has significant influence on dislocation dynamics for material design and mechanical properties of materials. Since dislocation interactions are the predominant deformation mechanisms that influence the mechanical property of coarse-grained materials, the plastic deformation mode of nanomaterials could be significantly different.
2. Due to the above reasons, alternate deformation mechanisms like grain boundary migration/sliding, crack growth, etc. become important.
3. The porosity level in nanomaterials is strongly dependent on the processing technique and can also be significant, for example, in the case of mechanically alloyed specimens. Even after agglomeration, the pore size can be smaller than or equal to grain size. The presence of porosity can have a significant effect on the mechanical behaviour of nanomaterials.
4. Finally, there can be segregation of different solutes at the grain boundaries in nanomaterials, which can influence the mechanical properties.

Some of the peculiar mechanical properties that were first observed for nanostructured materials are:

- Elastic moduli were found to be 30%–50% lower than in conventional grain size materials.
- Hardness and strength were found to be very high, with hardness values for nanocrystalline pure metals (~10 nm grain size) up to seven times higher than those for conventional coarse-grained (>1 μm) metals; in general, tensile and compressive strengths in nearly all materials show significantly high values at the nanometre scale.
- Nanoscale multilayers made of metallic or ceramic materials exhibited ultrahigh hardness.
- The Hall–Petch slope is negative below a critical grain size, showing that hardness decreases with decrease in grain size in the nanoscale grain size regime.
- Ductility (possibly superplastic behaviour) is observed at low temperature in brittle ceramics or intermetallics with nanoscale grain size, possibly due to diffusional deformation mechanisms–an area to be explored.

Hardness and strength

The hardness and strength of conventional grain size materials (grain diameter $d > 1$ μm) have been known to be a function of the grain size. It is known in general that grain boundaries act as barriers to dislocation motion and can thereby strengthen a material. With decreasing grain size, the surface area of grain boundaries per unit volume increases and therefore can result in an increase in hardness and strength. For ductile polycrystalline materials, a semi-empirical relationship, called the Hall–Petch equation, has been found to express the grain size dependence of the flow stress at any plastic strain, up to ductile fracture. In terms of yield stress, this expression can be written as

$$\sigma_y = \sigma_0 + k\, d^{-1/2}$$

where σ_y is the yield stress, σ_0 is a friction stress opposing dislocation motion, k is a constant and is material dependent, and d is the grain diameter. The same behaviour has been found to hold for hardness, as shown below, and for wear resistance too.

$$H = H_0 + k\, d^{-1/2}$$

This equation holds true for grains larger than about a micrometre. It is known that the yield stress–grain size exponent for relatively large grains appears to be very close to −0.5 and generally this trend continues until the very fine grain regime (~100 nm) is reached. For instance, for nanophase Cu (d = 6 nm), the hardness is five times greater than that in the corresponding bulk metal (d = 100 μm). This behaviour substantially derives from the difficulty in creating dislocations and from the existence of barriers to dislocation motion. Significant improvement in the tensile strength of nano-palladium in comparison to micrograined palladium has also been observed. In some cases, exceptionally high hardness has been observed in the nanocrystalline state, for example, in Al-based nanocomposites, prepared by mechanical alloying, as shown in Fig. 2.20. Copper-based nanocomposites have also shown very high hardness in comparison to micro-composies (Fig. 2.21). In fact, in these Cu-based nanocomposites, the ratio of hardness to electrical conductivity is three times that of oxygen-free high conductivity Cu-Aluminium-based nanocomposites also show retention of their hardness at higher temperatures, up to 773 K, in comparison to commercially available age hardened Al–Cu alloys, which lose their strength by 573 K (Fig. 2.22).

The mechanical behaviour is mainly determined by the type of bonding in the materials. Metals with highly delocalized bonds are ductile and are thus usually soft; ionic solids are more difficult to deform (due to charge neutrality); covalent solids have strong localised bonds, so that ceramics and intermetallics are subject to brittle fracture. In a crystalline structure, dislocations can be introduced and atomic planes can slide over one another. In the nanophase, both these behaviours are altered. In nanophase metals, the hardness increases with decreasing grain size because defects and dislocations are required for deformation to occur, and they are almost absent in nanostructured materials, although regenerative sources of dislocations exist. Therefore, the dislocations needed for easy deformation are frozen and

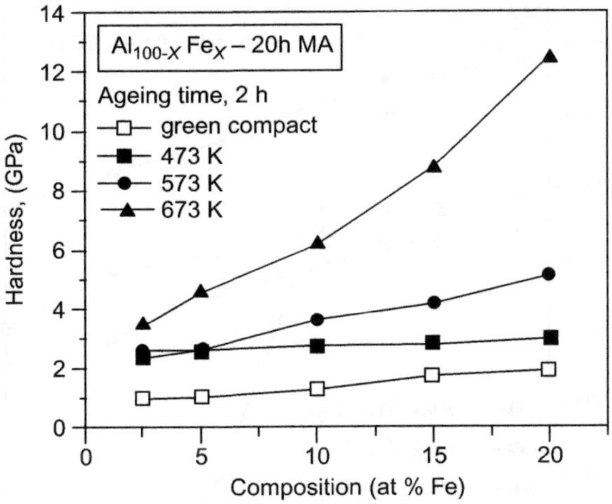

Fig. 2.20 Exceptionally high hardness of 12.4 GPa in Al-based nanocomposites. The nanocomposites contain $Al_{13}Fe_4$/quasicrystalline nanoparticles embedded in ultrafine Al grains. (*Source:* BS Murty, IIT Madras).

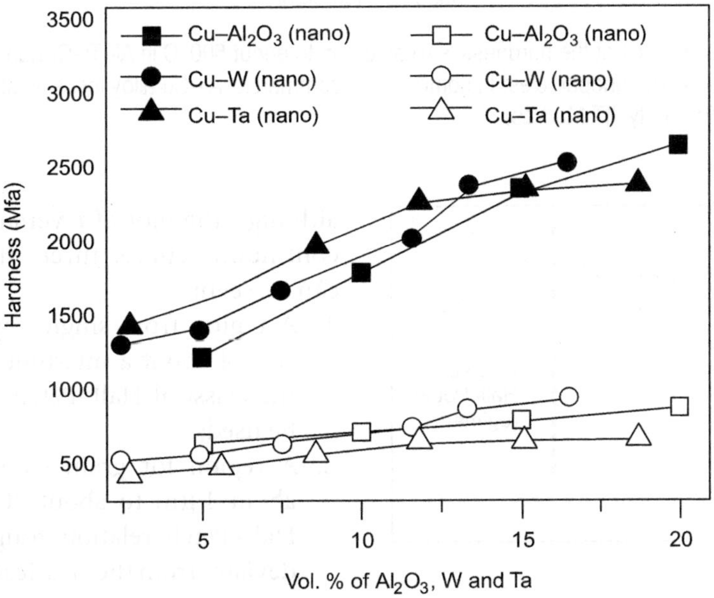

Fig. 2.21 High hardness of Cu-based nanocomposites in comparison to micro-composies. (*Source:* BS Murty, IIT Madras).

new ones are prevented from forming in nanomaterials.

However, in nanocrystalline materials, as the grain size is reduced below a critical limit, the response tends to deviate from the Hall–Petch relation described earlier (Fig. 2.23). In fact, a close analysis of experimental Hall–Petch data in a variety of materials has shown that

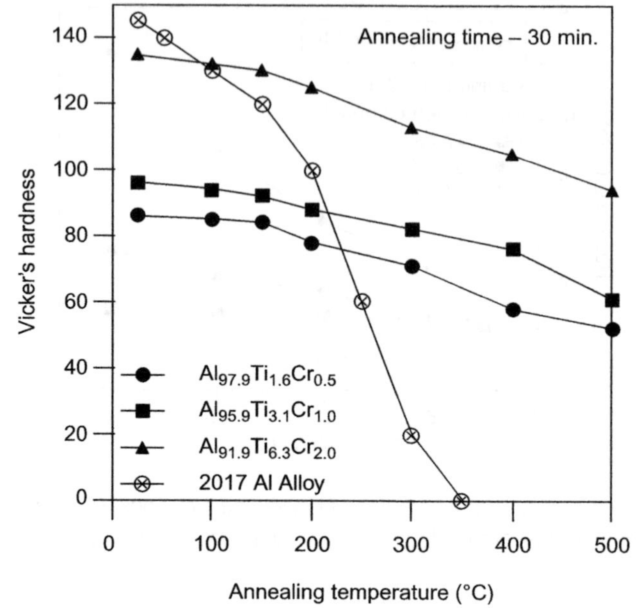

Fig. 2.22 The figure shows that the hardness is retained up to about 500°C in Al–Ti–Cr nanocomposites containing Al_3Ti intermetallic nanoparticles. In contrast, the commercial Al–Cu alloy loses its strength completely by 350°C. (*Source:* BS Murty, IIT Madras)

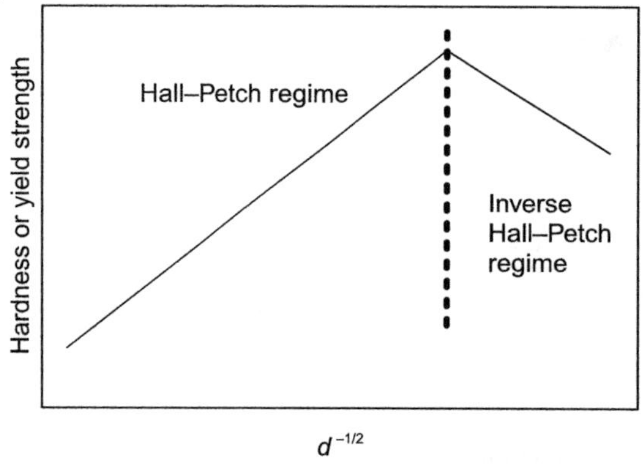

Fig. 2.23 Schematic figure showing inverse Hall–Petch effect below a critical grain size.

although the plot of τ versus $d^{-1/2}$ forms a continuous curve, three different regions can be seen:

1. A region from single crystal to a grain size of about a micrometre (μm) where the classical Hall–Petch description can be used;
2. A region for grain sizes ranging from about 1 μm to about 30 nm where the Hall–Petch relation roughly holds, but deviates from the classical – 0.5 exponent to a value near zero;
3. A region beyond a very small critical grain size where the Hall–Petch slope is nearly zero with no increase in strength on decreasing grain size or where the strength actually decreases with decreasing grain size.

It is known that below a critical grain size, the ratio of the volume fraction of triple junctions to grain boundaries increases significantly. As the triple junctions are known to be weak regions, it is expected that at this stage, the strength of the material decreases. Alternative explanations are discussed below. However, the field is under active investigation and a clear consensus is yet to be arrived at.

For large grain sizes, the dependence of stress can be rationalised in terms of a dislocation pile-up model. In deriving the Hall–Petch relation, the role of grain boundaries as a barrier to dislocation motion is considered in various models. In one type of model, the grain boundary acts as a barrier to pile-up of dislocations, causing stresses to concentrate and activating dislocation sources in the neighbouring grains, thus initiating slip from grain to grain. In the other type of model, the grain boundaries are regarded as dislocation barriers limiting the mean free path of the dislocations, thereby increasing strain hardening, resulting in a Hall–Petch-type relation. However, in nanocrystals, the number of dislocations decreases as described in previous sections. Lack of dislocation activity in the interior of the grains decreases the tendency for dislocation pile-up and thereby results in the deviation from the Hall–Petch relation.

Clearly, at sufficiently small grain sizes, the Hall–Petch model, based upon dislocations may not be operative, except possibly in the grain boundaries. It has been postulated that at very small grain sizes, a new mechanism of deformation may operate, which is similar to Coble creep or grain boundary diffusional creep occurring at room temperatures. Coble creep causes homogeneous elongation of grains in the tensile direction. Chokshi and his group at Indian Institute of Science, Bangalore, were the first to propose room temperature Coble creep as the mechanism to explain the so-called inverse Hall–Petch effect seen in nanocrystalline specimens. It should be noted that any artifacts of the specimen preparation methods can also contribute to negative Hall–Petch slopes and this is even more true for nanomaterials. One should therefore take great care during experimentation and interpretation of the mechanical properties of nanomaterials.

Tensile ductility and strain hardening

For microcrystalline materials, it is known that a decrease in grain size not only increases the strength but also the ductility. Grain size strengthening is perhaps the only mechanism that can enhance the hardness as well as ductility/toughness of a material. With decrease in grain size, crack deflections at grain junctions can result in enhanced constraint to crack propagation and thereby increase the fracture toughness and ductility of the material. When GB/triple junction migration is a more dominating mechanism of plastic deformation compared to dislocation, particularly as the grain size approaches close to the limit of instability of dislocations, it can result in early GB/triple junction crack interaction and crack propagation due to GB sliding even at low temperatures. When the grain size is reduced to nanocrystalline dimensions, one would expect an increase in the ductility of the material. However, the ductility of nanocrystalline materials is often observed to be reduced compared to ultrafine grained materials. Koch identified three major sources of limited ductility in nanocrystalline materials:

- artifacts from processing (e.g., pores)
- tensile instability
- crack nucleation or shear instability

Nanocrystalline bulk specimens synthesized for tensile testing usually contain a large fraction of porosities and other artifacts due to the synthesis technique. It is difficult to prepare bulk nanostructured materials free from such artifacts through consolidation. Theoretical and experimental studies suggest the existence of three regimes:

- Grain size $d > 1$ μm regime in which unit dislocations and work hardening control plasticity;
- Smallest grain size $d < 10$ nm regime, where limited intragranular dislocation activity occurs and grain boundary shear is believed to be the mechanism of deformation;
- Intermediate grain size regime (10 nm to 1 μm), which is less understood.

Nanocrystalline and ultrafine grained materials cannot generally sustain uniform tensile elongation. They do not show strain hardening after an initial stage of rapid strain hardening over a small plastic strain regime (1%–3%), which is different from the response of coarse-grained polycrystalline metal. This is because the dislocation density saturates in nanocrystalline materials due to both dynamic recovery as well as annihilation at grain boundaries. After large additional strains, work hardening is observed in these materials. Room temperature dynamic recovery is also common in nanocrystalline samples. An increase as well as a decrease in strain rate sensitivity with decreasing grain size in metals has been reported by different research groups. Iron, which is normally strain rate sensitive, with a strain rate exponent m of the order of 0.04, goes down in value to 0.004 when the grain size is 80 nm.

Creep and superplastic behaviour

If creep is determined mainly by conservative dislocation motion, i.e., motion along the glide planes of the dislocations, the creep behaviour of nanocrystalline materials is better than that of materials with conventional grain sizes. However, this occurs only at relatively high stress, close to the yield stress of the material, where the situation approaches that of a tensile test. At lower stress and higher temperatures, however, where the main mechanism governing creep is diffusion, creep generally occurs faster in materials with grain sizes in the nanometre range, i.e., where more high diffusivity paths are present.

The creep behaviour of nanocrystalline materials, however, cannot always be explained only on the basis of enhanced grain boundary diffusion. In fact, the creep rates of nanocrystalline materials are often lower than that predicted by the diffusion mechanisms along the grain boundaries alone. For example, doping a titanium oxide ceramic with yttrium oxide prevents grain growth during the test. According to a recent study, the creep rates of doped titanium oxide, at temperatures which are normally quite low for that material (973 and 1073 K), are lower than those of the non-doped one, i.e., with non-stabilised grains. At first glance, this is contrary to expectations.

Creep in nanophase materials is expected to occur quickly because of the larger volume fraction of grain boundaries and the short diffusion distances. Diffusion along the grain interfaces is rapid in nanophase metals, which promotes sintering and neck formation and hence enhanced creep rates are observed even close to room temperature. Creep rates may be influenced by the level of porosity in nanophase samples, because free surfaces tend to increase diffusion rates relative to grain boundary rates. Creep behaviour in nanophase ceramics can be understood using the Ashby–Verral model, which is based on the principle of grain boundary sliding with diffusional flow. If a compressive stress is applied to a system of equiaxed grains, there is diffusional flow in the boundary regions. Therefore, changes in the external shape occur; but the grains remain equiaxed and maintain their size and orientation, sliding over one another by diffusional accommodation at the interfaces. When a crack is opened at an interface, diffusion can contribute to fill it. The Karch model relates the strain/creep rate to grain boundary diffusion

$$\frac{d\varepsilon}{dt} = D \frac{\sigma\Omega}{kT} \left(\frac{b}{d}\right)^3$$

where σ is the applied stress, Ω is the atomic volume, d is the grain size, b is Burger's vector, k is the Boltzmann constant, and T is the grain boundary diffusion coefficient. When the grain size is decreased from 1 μm to 10 nm, $d\varepsilon/dt$ is increased by 10^6 or more. Eventually, exceptional ductility due to diffusional creep in nanocrystalline brittle ceramics or intermetallics at temperatures less than 0.5 T_m has not been realised, while effective enhanced ductility has been obtained at somewhat higher temperatures, and an improvement of creep properties by lower 'activation' temperature for superplasticity and a higher strain rate regime.

The ductile–brittle transition temperature in conventional mild steel can be lowered by about 300 K by reducing the grain size by a factor of 5. The brittleness can be defined in terms of the strain rate sensitivity, m, as given below, and its values are 0 (perfectly brittle) and 1 (perfectly ductile).

$$\sigma = k \left(\frac{d\varepsilon}{dt}\right)^m$$

Some polycrystalline materials are able to exhibit very large tensile deformations without necking or fracture. Elongations of 100%–1000% are considered the defining feature of this phenomenon. As grain size decreases, it is found that the temperature at which superplasticity occurs is lowered, and the strain rate for its occurrence is increased. Superplastic behaviour might be observed in nanocrystalline materials at temperatures much lower than $0.5T_m$, but instead have shown creep rates comparable to or lower than those in coarse-grained samples of the same material. Therefore, a little enhancement in ductility or superplastic behaviour has been observed for nanocrystalline materials at temperatures $<0.5T_m$. This provided evidence for the enhancement of superplastic behaviour in nanocrystalline materials at temperatures $>0.5T_m$. Superplasticity has been observed at somewhat lower temperatures and at higher strain rates in nanocrystalline materials. The evidence for tensile superplasticity is limited and observed typically at temperatures greater than $0.5T_m$ and in materials that exhibit superplasticity in coarser grain sizes (1–10 μm).

Superplasticity means the capability of polycrystalline materials to exhibit very large tensile deformations without showing fracture or necking. It is observed only above a certain critical temperature in the materials that present this behaviour and also when they are coarse-grained. These 'activation' temperatures are usually higher than $0.5T_m$, where T_m is the melting temperature. When the grain size is decreased, it is found that the threshold temperature can be lowered; an increase in the strain rate for the occurrence of superplasticity is observed. It was believed that creep rates could be enhanced by many orders of magnitude by reducing the grain size below 100 nm, and that superplastic behaviour could be observed in nanocrystalline materials at temperatures much below $0.5T_m$. Superplasticity has been observed at lower temperatures and at higher strain rates in nanocrystalline materials than in coarse-grained materials. It must be observed that tensile superplasticity is limited to materials that already exhibit superplasticity in coarser grain sizes (1–10 μm), e.g., for ZrO_2 and TiO_2, $m = 0.02$ at 300 K for $d = 100$ nm, but m increases almost exponentially below 50 nm for both the oxides.

Fracture and toughness

In practical applications of materials, a significant limitation is represented by the fracture strength, and not by the yield strength. A material can yield and with this mechanism sustain itself against an imposed stress. This is possible because of behaviour dislocations and grain boundaries, which prolong plasticity. However, a material possessing elevated strength can also fail suddenly in a brittle way, with very little plasticity. This is a problem for applications in the real world.

The results of the bending test for compacted NiP showed that by decreasing the grain size, an enhancement of the fracture stress and the strain-to-fracture can be realised. For microstructured materials, improvements in toughness and a lower brittle-to-ductile transition temperature are obtained by decreasing the grain size. This is also true for nanomaterials. However, in brittle intermetallics such as TiAl (grain size ~ 100 nm to 70 μm), the toughness was found to decrease with decreasing grain size, as intragranular cracking gave way to intergranular fracture. Mechanical failure, which limits ductility, can be viewed as a competition between dislocations and nucleation or propagation of cracks, depending on the grain size. However, the large increase in yield stress observed in nanomaterials suggests that fracture stress can be lower than the yield stress and therefore result in nil ductility.

Corrosion properties

Several reports suggest that the high density of grain boundaries in nanocrystalline alloys is responsible for their significantly changed properties. The corrosion investigations of nanocrystalline materials showed a more intensive active anodic dissolution, compared with conventional materials. This difference in behaviour can be explained by the small grain size, which results in a larger fraction of high-energy grain boundary defects in the material. Such high-energy grain boundary sites act as preferred anodic dissolution sites and can thus result in increased corrosion rates of nanocrystalline alloys. However, higher defect densities are not always disadvantageous. Grain boundaries being easy diffusion paths can facilitate enhanced surface diffusion coefficients, resulting in the early formation of a well-adherent and dense oxide layer, thereby passivating the surface. Some authors showed that the grain size does not affect the thickness of the passive film. Electrochemical polarization tests on nanocrystalline Ni (grain sizes of 500 and 20 nm) indicated that they exhibit higher passive current densities than their coarser-grained conventional Ni counterparts.

XPS investigations of the passive film on these materials showed no difference in the thickness of the passive film between the nanocrystalline material and conventional material. These passive films on the nanocrystalline specimens were, however, more defective than those on conventional Ni. The defect density of the film was found to increase with decreasing grain size, which in turn allowed for easier Ni cation diffusion through more defective film, leading to higher current densities in the passive potential range for the nanocrystalline specimens. It was also observed that when the grain size is in the nanoscale range, the number of atoms at the grain boundaries becomes comparable to the number of atoms inside the grains. The amount of impurities per unit grain boundary is therefore lower in nanomaterials than larger-grained materials with the same bulk impurity concentration, because the grain boundary area increases. This higher purity of the grain boundaries has been associated with more uniform corrosion morphology and higher intergranular corrosion resistance for nanomaterials as compared to larger-grained materials. One more factor which affects the corrosion behaviour of nanocrystalline material is the porosity of the coating. If the coating is less porous in nature, it will show high corrosion resistance. Contribution of triple junctions to the structure and properties of nanocrystalline materials may also affect the corrosion properties of these nanocrystalline materials (Table 2.4). Copper-based nanocomposites show significantly better corrosion resistance when compared to the microcomposites, as shown in Table 2.5.

Table 2.4 Corrosion behaviour of nanomaterials

Characteristic	Beneficial effects	Deleterious effects
Enhanced diffusion	Faster surface diffusion of ions to enhance kinetics of passive film formation	Faster diffusion can result in higher corrosion currents
Stability of passive film	Faster grain boundary diffusion can aid healing of passive film in service	Porous oxide film
Effect of large grain boundary area	Lower dilution effects	Increased anodic sites for nucleation of corrosion damage
High temperature effect	Better thermal shock resistance	Rapid grain growth limits high temperature application

Table 2.5 Corrosion resistance of Cu-based nanocomposites in relation to micro-composies

Vol.% / Wt.% of dispersoids	Corrosion rate (mpy) Micro	Nano	Fraction decrease wrt. microcomposite
$Cu–Al_2O_3$			
5.0/2.3	76	17	4.4
10.0 /4.7	176	77	2.3
20.0/10.0	288	130	2.2
Cu–W			
2.4/5.0	90	5	18
11.6/20.0	122	7	17
19.8/30.0	246	33	7
Cu–Ta			
2.8/5.0	80	4	20
13.3/20.0	99	5	20
22.9/30.0	141	7	20

SUMMARY

- The unique properties of nanomaterials arise from their small size and structural features such as large fraction of grain boundaries, triple junctions, etc.
- Initially, it was thought that the grain boundaries in nanomaterials are 'gas-like'; recently, there has been significant improvement in the understanding of the nature of grain boundaries in these materials.
- A number of unanswered questions remain on the extent of defects such as dislocations, disclinations, etc., which influence the properties of these materials significantly.
- The processing routes adopted to synthesize nanomaterials can significantly influence the grain boundary structure and the concentration of other defects in these materials, which in turn can influence their properties.

EXERCISES

1. Based on thermodynamic considerations, explain the reasons for the difference in the melting point of a free-standing nanoparticle of a pure metal and an embedded nanoparticle, in comparison to the bulk metal.
2. What is the principle behind exchange-coupled magnets?
3. What is the GMR effect?
4. Discuss the stability of dislocations and disclinations in nanocrystalline materials.
5. Calculate the activation energies for various diffusion paths shown in Fig. 2.9. Please indicate your assumptions during the calculation. Discuss the reasons for the differences in activation energies.

6. Explain the principle of Zener pinning and solute drag effects during grain growth in nanomaterials.
7. Write a short note on thermal conductivity of nanofluids.
8. Describe the inverse Hall–Petch relation.
9. The hardness of Fe is 4 and 6 GPa at 80 and 60 nm, respectively. Assuming that the Hall–Petch relation is valid, calculate the hardness of Fe with a grain size of 7 nm.
10. Explain the reasons for improved soft magnetic properties in nanocrystalline magnets.

Herbert Gleiter received his PhD from the Department of Physics, University of Stuttgart, Germany, in 1966, and worked as a professor in the Departments of Materials at Bochum University and Saarland University, successively. He is a faculty member at both Hamburg University and ETH Zurich. Professor Gleiter conducted research as Visiting Professor at Technion University, at Bell Laboratories, University of Wisconsin, at Monash University, at Tohoku University, at Beijing University and at the University of New South Wales. In 1987, he founded the Institute of New Materials at Saarland University, at which he worked as a director. In 1994, he was appointed a member of the Executive Committee of the Karlsruhe Research Centre in Germany. From 1998, he has been with Karlsruhe Institute of Nano-Materials, which was founded by him together with Nobel Prize Winner Prof. Jean-Marie Lehn and Prof. Fenske. It is a premier institute in nanomaterials technology research in Germany.

Twenty years ago, Professor Gleiter put forward a new idea for the structure of nanomaterials, and initiated a new realm of materials science research. He is an academician of Germany Leopoldina Academy of Science, and a foreign member of the US Academies of Sciences and Engineering. Prof. Gleiter has published more than 350 papers on Nature and Science and seven monographs. He has been awarded the Gold Medal by the European Association of Materials, the Leibniz Prize by the Germany National Foundation of Science, the Vinci Excellence Award by the Hennessy Vuitton Foundation, the Max Planck Research Prize by the Alexander von Humboldt Foundation, the Heyn Prize by the German Institute of Material Science, the TMS Prize by the American Institute of Mineral Metal and Material, the Annual Prize by the Japan Institute of Metals, and several others.

Chapter 3

Synthesis Routes

Learning objectives

- Different routes for the synthesis of nanoparticles and nanocrystalline materials
- Different routes for the consolidation of nanoparticles and nanocrystalline materials

There are different ways of classifying the synthesis routes for nanostructured materials. One of them is based on the starting state of material, namely, gas, liquid and solid. Techniques such as vapour condensation [physical vapour deposition (PVD) and chemical vapour deposition (CVD) and variants of these techniques] use the gaseous state of matter as the starting material for synthesizing nanoparticles. Techniques such as sol–gel, chemical and electrochemical (electrolytic) deposition and rapid solidification processing use liquids as the starting material. Severe plastic deformation processes such as high-energy ball milling, equichannel angular extrusion, etc., and nano-lithography, start with solids for synthesizing nanocrystalline materials.

However, the most popular way of classifying the synthesis routes is based on how the nanostructures are built, and such an approach leads to two routes, namely, the 'bottom-up' and the 'top-down' approaches. In the bottom-up approach, individual atoms and molecules are brought together or self-assembled to form nanostructured materials in at least one dimension. All the techniques that start with liquid and gas as the starting material fall into this category. In the second approach (top-down approach), a microcrystalline material is fragmented to yield a nanocrystalline material. All the solid state routes fall into this category.

Usually, the bottom-up techniques can give very fine nanostructures of individual nanoparticles, nanoshells, etc., with narrow size distributions, if the process parameters are effectively controlled. The top-down techniques do not usually lead to individual nanoparticles; however, they can produce bulk nanostructured materials. Many of the bottom-up approaches have difficulties in scale up, while the top-down approaches can be easily scaled up. Thus, one can see that both these approaches are complementary to each other, depending on the requirement of a particular application. The most prominent techniques to synthesize nanostructured materials are described.

3.1 BOTTOM-UP APPROACHES

3.1.1 Physical vapour deposition (PVD)

PVD is a versatile synthesis method and is capable of preparing thin film materials with control at the nanometre scale by careful monitoring of the processing conditions. PVD involves the generation of vapour phase either via evaporation, sputtering, laser ablation or by using an ion beam. In *evaporation*, atoms are removed from the source, usually by heating it above its melting point. On the other hand, in *sputtering*, atoms are ejected from the target surface by the impact of energetic ions. Thermal evaporation has a limitation in multicomponent materials since one of the metallic elements typically evaporates before the other, due to the differences in boiling point and vapour pressure of the evaporating species. On the contrary, sputtering is capable of depositing high melting point materials such as refractory metals and ceramics, which are difficult to convert to nanomaterials by evaporation. Sputtering can result in better stoichiometric control of the film compared to evaporation techniques. Sputter-grown films usually have higher density than those obtained by evaporation as the sputtered atoms have more energy than the evaporated atoms. Sputtered films are more prone to contamination than evaporated films due to the lower purity of the sputtering target materials.

INERT GAS CONDENSATION
(IGC) combined with thermal evaporation is commonly used to synthesise metallic and metal oxide nanopowders with a well-defined and narrow size distribution. This technique was originally introduced by Ganqvist and Buhrman in 1976 and later developed by Gleiter in 1981. In this process, a metal is evaporated inside an ultra-high vacuum (UHV) chamber filled with inert gas, typically helium (Fig. 3.1). The vapourised species then loses energy

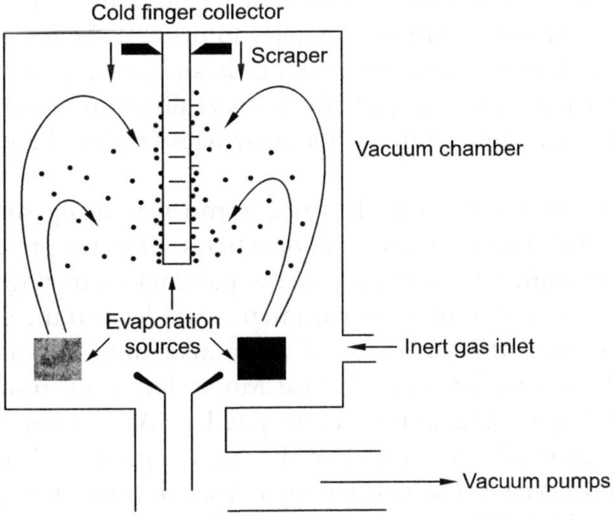

Fig. 3.1 Schematic inert gas condensation unit for the synthesis of nanocrystalline particles.

via collisions with helium molecules. As collisions limit the mean free path, supersaturation can be achieved above the vapour source. At high supersaturation, the vapours rapidly form large numbers of clusters that grow via coalescence and agglomeration. The clusters in the condensing gas are transported by convective flow to a vertical cold finger surface filled with liquid nitrogen. The removal of particles from the cold finger is carried out by a scraper assembly. They are collected via a funnel and transported to an in situ compaction device or coated with a surfactant agent that prevents them from agglomeration. The scraping and consolidation of particles are carried out under UHV conditions to prevent oxidation of the metallic nanoparticles. The size, morphology and yield of the clusters in gas condensation are dependent on three fundamental parameters:

- Rate of supply of atoms to the region of supersaturation where condensation occurs
- Rate of energy removal from the hot atoms via the condensing gas medium
- Rate of removal of clusters once nucleated from the supersaturated region

Particle nucleation, coalescence and growth during condensation also play key roles in forming small particles in large numbers.

As the yield of inert gas deposition is rather low, Bigot and co-workers developed a process wherein the metallic vapours condensed into a cryogenic medium to produce metallic nanocrystals. The main advantage of this technique is that it exhibits a higher production rate of about 60 g/h and a yield of 75%. Rapid overheating of the metal via radio frequency (RF) induction technique produces a high vapour pressure and substantial evaporation rate. The reactor is supplied continuously with cryogenic liquid, and nanoparticles are formed by rapid condensation of the supersaturated metal vapour. The condensation region, where the particles are formed by the nucleation, growth and coalescence processes, features a high temperature gradient—typically from 2200 K at the metallic surface to 77 K in the cryogenic medium. The low temperature of the surrounding medium produces a high rate of nucleation and rapid cooling of the as-formed particles limits crystal growth. Liquid argon is used for the synthesis of Al nanoparticles to prevent the formation of aluminium nitrides. Liquid nitrogen has been used for making Cu and Fe nanopowders. This technique yields Cu nanoparticles of spherical shape with their size distribution following a log-normal function peaking at 25 nm. The aluminium nanoparticles produced are also spherical, with size less than 70 nm.

IGC using a direct current (DC) or RF magnetron sputtering source has been used to prepare refractory metal and ceramic nanoparticles. The gas pressure, sputter power and source substrate distance are crucial process parameters during sputtering and gas condensation. A careful selection of these parameters will lead to the successful formation of nanoparticles rather than the formation of a granular film. The reported particle size is strongly dependent on argon pressure and for Mo, it decreases from 12 nm at 0.4 mbar to 5 nm at 0.8 mbar. Figure 3.2a shows nanocrystalline Ag–Cu deposit obtained by the IGC technique. The nanocrystalline nature of the Ag–Cu particles is clearly evident from the diffraction rings observed in the selected area electron diffraction pattern of Fig. 3.2b. Figures 3.2c–3.2f show the TEM micrographs of the deposit heated in situ to 200, 300, 400 and 600°C, respectively, in the TEM for understanding the grain growth behaviour of the

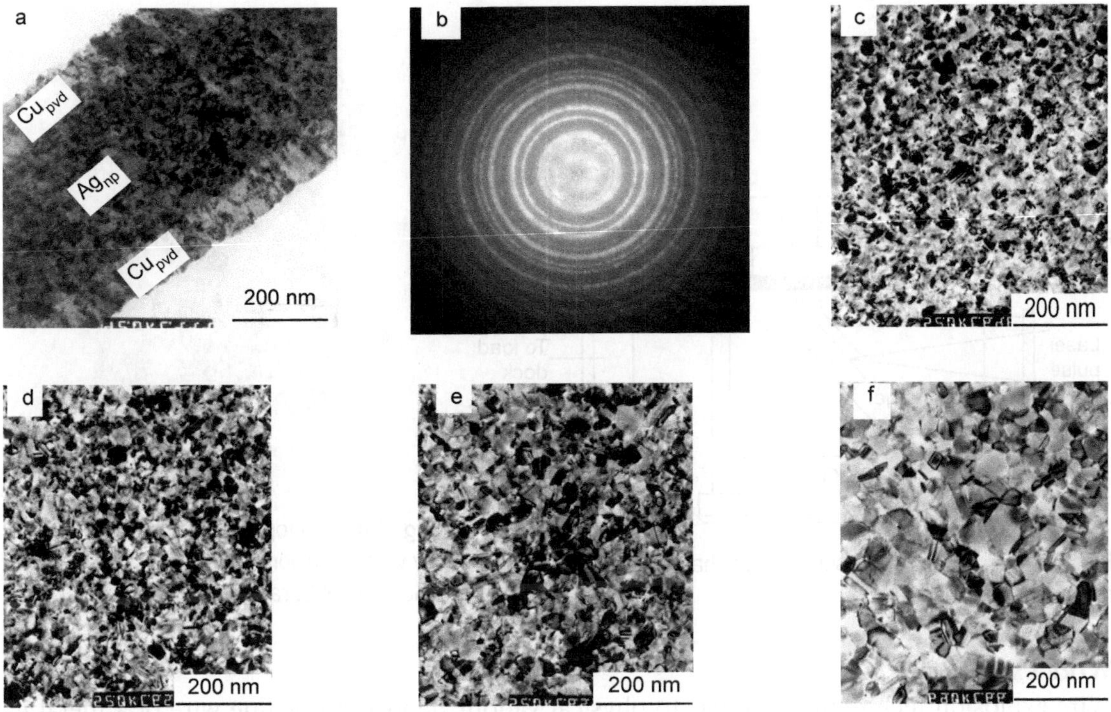

Fig. 3.2 Ag–Cu nanoparticles—(a) and (b)—synthesised by IGC and subjected to heating to (c) 200, (d) 300, (e) 400 and (f) 600°C. (*Source:* BS Murty, IIT Madras).

nanocrystalline deposit. The figure clearly indicates that nanocrystalline nature is retained in the deposit even after heating it to 600°C, due to the solute drag effect from the Cu present in Ag.

Laser ablation

It is generally known that laser ablation can provide better control of the evaporation process by congruent evaporation of constituent elements of multicomponent materials in a very short period of time. In this technique, an intense pulsed laser beam irradiates the target of interest, thereby vaporising atoms and clusters from the target. The total mass ablated from the target per laser pulse is usually referred to as the ablation rate. Laser ablation in combination with IGC is an attractive route to synthesize larger amounts of the multicomponent nanocrystalline materials. A typical laser ablation set up is shown in Fig. 3.3. In the process, atoms ablated by laser pulse tend to collide with helium gas, losing their kinetic energy rapidly. They eventually condense to form a cloud consisting of fine nanocrystalline clusters. The production rate of nanoparticles in this technique varies with helium gas pressure and laser pulse energy. Several workers have employed laser ablation and gas condensation to produce nanoparticles of metals, metal oxides and metal carbides.

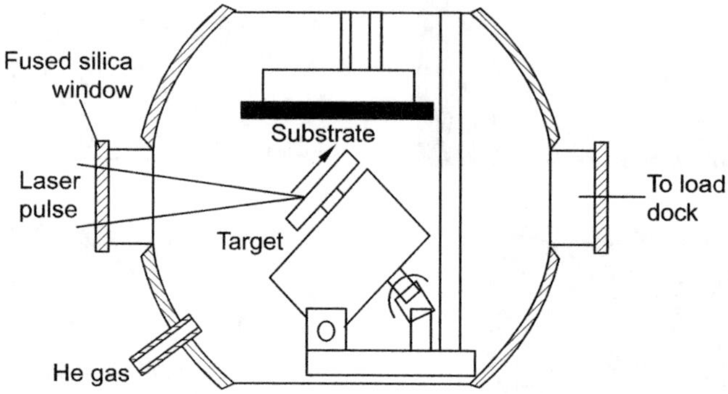

Fig. 3.3 Schematic of a laser ablation chamber equipped with a rotating target holder.

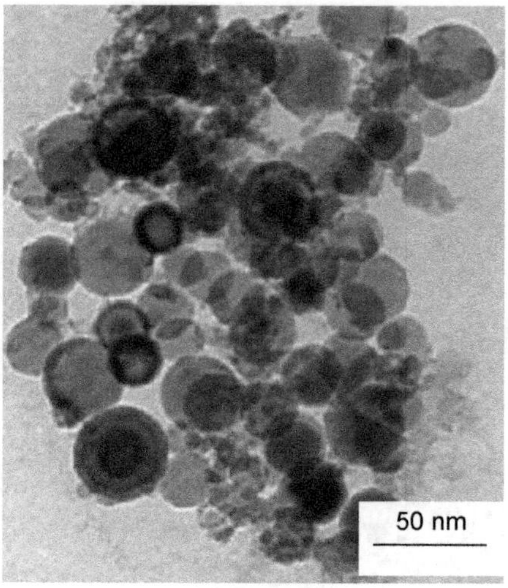

Fig. 3.4 Nanoparticles of Al synthesized by wire explosion technique.
(*Source:* R Sarathi, IIT Madras)

WIRE EXPLOSION

Wire explosion technique is similar to physical vapour deposition. A fine wire of a metal is exploded by applying a very high voltage. The gas of atoms generated by this explosion is allowed to condense in the chamber to yield nanoparticles. This technique can lead to the formation of not only metallic nanoparticles but also a variety of oxides, nitrides, etc. by using different environments in the chamber. A typical example of nanoparticles generated by this technique is shown in Fig. 3.4.

3.1.2 Chemical vapour deposition

Chemical vapour deposition (CVD) is a process where one or more gaseous adsorption species react or decompose on a hot surface to form stable solid products. The main steps that occur in the CVD process can be summarized as follows:

1. Transport of reacting gaseous species to the surface
2. Adsorption of the species on the surface
3. Heterogeneous surface reaction catalysed by the surface
4. Surface diffusion of the species to growth sites
5. Nucleation and growth of the film
6. Desorption of gaseous reaction products and transportation of reaction products away from the surface

CVD is a more complicated method than PVD for the formation of thin films and coatings. It exhibits several distinct advantages, such as the capability to produce highly pure and dense films or fine particles at reasonably high deposition rates, and the capability of coating complex-shaped components uniformly due to its non–line-of-sight nature. A variety of metallic, ceramic and semiconducting thin films are being deposited by CVD. Depending on the activation sources for the chemical reactions, the deposition process can be categorized into thermally activated, laser-assisted and plasma-assisted CVD.

Thermally activated CVD

This is the conventional CVD technique where the resistive heating of hot wall reactors gives sufficiently high temperatures for the dissociation of gaseous species. This leads to the heating of the entire substrate to a high temperature before the desired reaction is achieved. It precludes the use of substrates with melting points much lower than the reaction temperature. Hot tungsten filament is also used to heat the reacting gases in the vicinity of the substrate. Recently, hot filament CVD was used to grow one-dimensional SiC nanorods. Due to high hardness, high thermal stability, wide band gap and high electron mobility, SiC nanorods are potential materials for structural and electronic applications. Further, the Young's modulus of the nanorod can reach ~ 600 GPa, which is close to the theoretically predicted value for [111]-oriented SiC. In this CVD process, a silicon wafer was initially immersed in aqueous $Fe(NO_3)_3$ solution to remove the contaminants before transferring it into the reactor chamber. Carbon, silicon and silicon dioxide powders, compacted in the form of a plate, are used as the silicon and carbon sources. The chamber was filled with hydrogen gas and a tungsten filament was used to activate the hydrogen gas. The filament and substrate temperatures used are 2570 and 1373 K, respectively. The hot filament causes dissociation of molecular hydrogen to atomic hydrogen. The $Fe(NO_3)_3$ particles on the Si wafer surface were reduced by high concentration of atomic hydrogen to iron nanoparticles, which acted as a catalyst. In order to form hydrocarbon and silicon monoxide radicals, the silicon and carbon sources are thermally activated by hot filaments. The rods are straight with diameters of 20–70 nm and lengths of about 1 mm. Iron catalyst particles can be seen at the tips of the nanorods. HRTEM images revealed that the nanorods grew along the [100] direction. An electron bright field image of gold nanorods is shown in Fig. 3.5.

Plasma-enhanced CVD

This has a distinct advantage over thermal CVD due to its lower deposition temperature. The plasma is generated by various energy resources such as DC, RF, microwave and electron cyclotron resonance microwave (ECR-MW) radiation. In DC plasma, the reacting gases are ionised and dissociated by an electrical discharge, generating plasma consisting of electrons and ions. Microwave plasma is attractive because the excitation microwave frequency (2.45 GHz) can oscillate electrons. Thus, high ionization fractions are generated as electrons collide with gas atoms and molecules. SiC films have been deposited on silicon substrates by RF plasma-enhanced CVD using a gaseous mixture of $SiCl_4$, CH_4, H_2 and Ar. Recently, microwave plasma-enhanced CVD (MW-CVD) has been used to grow one-dimensional carbon nanotubes at a low temperature range of 600–800°C. Cold cathodes in field emission displays need well-aligned carbon nanotube arrays. The MW-CVD method has the advantage of achieving good alignment and selective growth on various substrates.

Conventional CVD of hydrocarbons over metal catalyst has been a typical method to produce various forms of carbon fibres, filaments and multiwalled nanotubes. In this process,

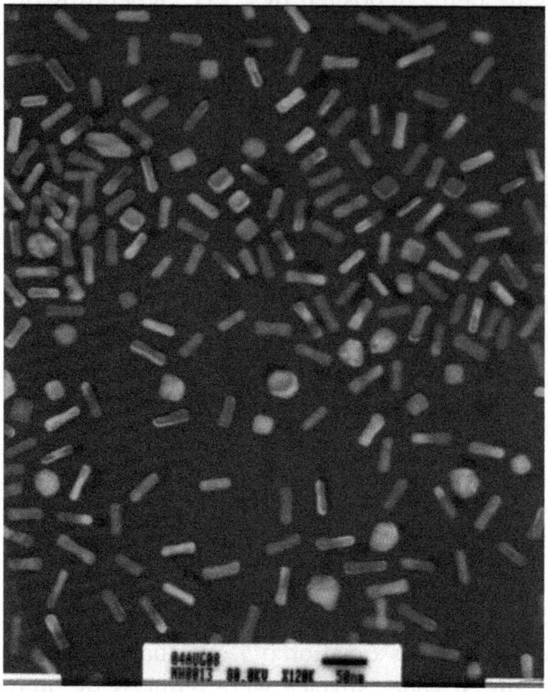

Fig. 3.5 Transmission electron micrograph of gold nanorods, along with a small fraction of impurities. (*Source:* http://commons.wikimedia.org/wiki/File: Au_nr_mhore.jpg.).

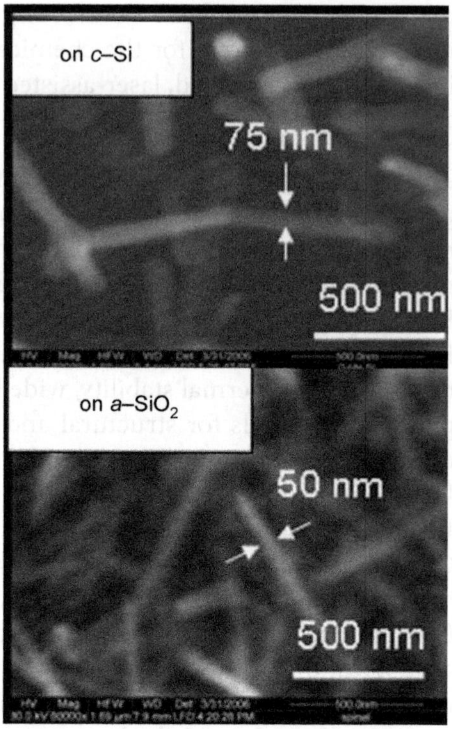

Fig. 3.6 GaN nanowires deposited by CVD on crystalline Si and amorphous silica. (*Source:* P Shankar, Saveetha University).

hydrocarbon molecules are absorbed and decomposed on transition metal (Fe, Ni, Co) particles. The carbon atoms diffuse into the interior of the catalyst to form a metal–carbon solid-state reaction. The precipitation of carbon from the supersaturated catalyst particle gives rise to carbon tube structure. Figure 3.6 shows GaN nanowires deposited by CVD on crystalline Si and amorphous silica.

Laser CVD is associated with the deposition of chemical vapours using a laser beam generated from sources like CO_2, Nd:YAG and Excimer. This process yields good quality films at lower temperatures with better control on composition. For example, a silicon nitride film could be deposited at 200°C using laser CVD, whereas it is deposited at 850°C and 450°C by thermally activated CVD and plasma-enhanced CVD, respectively. In the case of nanoparticles, tungsten powder of 54 nm could be synthesised from $WF_6/H_2/M$ (M = Ar, Kr, Ne, Xe) gas mixtures irradiated with an ArF Excimer laser. Moreover, ceramic Si–C–N nanometric powders can be obtained from laser-induced gas phase reactions of SiH_4, amines and NH_3. The incorporation of Si–C–N powders with sizes of 25–30 nm into aluminum improves its tensile and creep properties considerably. The box below describes a technique (molecular beam epitaxy) for precise formation of nanostructured thin films. There have also been attempts by people to

MOLECULAR BEAM EPITAXY (MBE)

MBE is a special case of evaporation for single crystal film growth, with highly controlled evaporation of a variety of sources in ultrahigh vacuum of about 10^{-10} torr. Besides the ultrahigh vacuum system, the MBE system usually consists of real-time structural and chemical characterisation capability, including reflection high energy electron diffraction (RHEED), x-ray photoelectric spectroscopy (XPS) and Auger electron spectroscopy (AES). Other analytic instruments may also be attached to the deposition chamber or to a separate analytic chamber, from which the grown films can be transferred to and from the growth chamber without exposure to the ambient atmosphere.

In MBE, the evaporated atoms or molecules from one or more sources do not interact with each other in the vapour phase under such low pressure. Although some gaseous sources are used in MBE, most molecular beams are generated by heating solid materials placed in source cells, which are referred to as *effusion cells* or *Knudsen cells*. The source materials are most commonly raised to the desired temperatures by resistive heating. The mean free path of atoms or molecules (100 m) far exceeds the distance between the source and the substrate (typically 30 cm) inside the deposition chamber. The atoms or molecules striking the single crystal substrate result in the formation of the desired epitaxial film. The extremely clean environment, the slow growth rate and independent control of the evaporation of individual sources enable the precise fabrication of nanostructures and nanomaterials as a single atomic layer. The ultrahigh vacuum environment ensures the absence of impurity or contamination, and thus a highly pure film can be readily obtained. Individually controlled evaporation of sources permits the precise control of chemical composition of the deposit at any given time. The slow growth rate ensures sufficient surface diffusion and relaxation so that the formation of any crystal defect is kept minimal.

ATOM MANIPULATION

In addition to imaging surface topography at atomic resolution, scanning tunnelling microscopy (STM) has the ability to carry out precise and controlled manipulation of atoms, molecules and nanostructures. The usual technique to manipulate atoms is to increase the current above a certain atom, which reduces the tip–atom distance, move the tip with the atom to a desired position, and finally to reduce the current again in order to decouple the atom and tip. The first demonstration of this technique was performed by Eigler and Schweizer (1990), who used Xe atoms on an Ni (110) surface to write the three letters 'IBM' on the atomic scale. Currently, many laboratories are able to move different kinds of atoms and molecules on different surfaces with high precision. Controlled motion, pushing, pulling and sliding of the molecules depends on the tunnelling current, the distance and the particular combination of molecule and substrate. It is believed that the electric field between tip and molecule is the strongest force moving the molecules, but other mechanisms such as electromigration caused by the high current density or modifications of the surface potential due to the presence of the tip may also play an important role.

manipulate atoms and arrange them 'the way we want' as has been dreamt by Feynman. These efforts are described in the box above (atom manipulation).

3.1.3 Spray conversion processing

This route involves the atomization of chemical precursors into aerosol droplets that are dispersed throughout a gas medium. The aerosols are then transported into a heated reactor where the solution is evaporated to form ultrafine particles or thin films. This is an inexpensive technique as various low cost chemical solutions are available. Various aerosol generators—including pressure, electrostatic and ultrasonic atomisers—have been used for atomization purposes. These atomizers affect the droplet size, rate of atomization and droplet velocity. The most commonly used aerosol processing method is *spray pyrolysis*. In the process, aqueous solution is atomized in a series of reactors where the aerosol droplets undergo evaporation and solute condensation within the droplet; drying and thermolysis is followed by sintering. Nanoparticles can be prepared directly, synthesized from droplets or by liberating individual crystallites comprising the spray pyrolysis-derived particles from the thermolysis stage. Spray pyrolysis can be used to prepare several metal oxide nanoparticles such as ZnO, ZrO_2 and Al_2O_3.

Combustion flame spraying

Another type of gas condensation technique that uses liquid chemical precursors as the starting material is combustion flame spraying. The advantages of this process are low cost and high production rates. In the combustion flame spraying process, burning a fuel–oxygen mixture with a spray torch or flat-flame burner generates a flame. Chemical precursors are vapourised in the hot zone of the flame and pyrolysis occurs in the thin hot zone. The interaction between the flame and droplets yields nanoparticles. The use of a low-pressure chamber for the flat-flame burner precludes contamination and ensures the temperature profile across the entire burner surface is uniform. Low-pressure combustion flame spraying is often referred to as combustion flame chemical vapour condensation (CF-CVC). Solid, liquid and gaseous precursors can be used in this technique. CVC is an alternative to the inert gas phase condensation approach in which the evaporator in the IGC system is replaced by other sources such as burning flame, hot wall furnace and microwave plasma. The major advantage of plasma-assisted pyrolysis in contrast to thermal activation is the low reaction temperature which reduces the tendency for agglomeration of nanoparticles. Combustion flame spraying has been used to produce a variety of high purity and non-agglomerated metal oxide nanoparticles such as TiO_2, Al_2O_3, ZrO_2, V_2O_5 and Y_2O_3–ZrO_2.

3.1.4 Sol–gel process

The sol–gel processing method has been in use for many years for producing metal oxide and ceramic powders with high purity and high homogeneity. The sol–gel route offers a degree of control of composition and structure at the molecular level. The process involves the generation of a colloidal suspension ('sol'), which is subsequently converted to viscous gel and solid material. Ebelman produced the first silica gel in 1846, and Cossa synthesized alumina gel in 1870. Since then, aerogels of zirconia, silazane, borate and other ceramics have been synthesized using the sol–gel technique.

In the process, reactive metal precursors were initially hydrolysed, followed by condensation and polymerization reactions. Metal alkoxides are metalorganic compounds with an organic ligand attached to a metal or metallic atom. They are the result of direct or indirect reactions between a metal M and an alcohol ROH. Typical examples are methoxide (OMe; $MOCH_3$) and ethoxide (OEt; MOC_2H_5). During hydrolysis, the alkoxy group OR is replaced by hydroxo ligands (OH), i.e.,

$$M(OR)_z + H_2O \longrightarrow M(OH)(OR)_{z-1} + ROH$$

where R is an alkyl group, C_nH_{2n+1}. The mechanism of this reaction involves the addition of a negatively charged $HO^{\delta-}$ group to the positively charged metal centre ($M^{\delta+}$) followed by the removal of ROH.

The factors that influence the hydrolysis reaction are:

- Nature of the alkyl group
- Nature of the solvent
- Concentration of each species in the solvent
- Temperature
- Water to alkoxide molar ratio
- Presence of acid or base catalysts

Subsequent condensation eliminates either water or alcohol to produce metal oxide or hydroxide linkages. In this process, two mononuclear complexes of M, each comprising only one metal M, can react with one another to form a polynuclear complex consisting of two metal atoms. Condensation occurs only when at least one hydroxo ligand is bonded to the cation M, and is designated as M-OH for simplicity. Olation and oxolation reactions lead to condensation. Olation is a reaction by which the hydroxo or 'ol' bridge (M–OH–M) is formed between two cations, while oxolation involves the formation of oxo bridges (M–O–M) between two metal cations:

$$\text{olation:} \quad M - OH + M - OH_2 \longrightarrow M - OH - M + H_2O$$

$$\text{oxolation:}\quad M - OH + H - OM \longrightarrow M - O - M + H_2O$$

$$M - OH + ROM \longrightarrow M - O - M + ROH$$

The 'ol' or 'oxo' bridges between two metal atoms lead to the formation of condensed oxide or hydroxide species. Under acid conditions, three-dimensional solid phase networks consisting of extended linear M–O–M chain polymers are developed. Inorganic polymerization is believed to occur in three stages during acid-catalysed condensation:

1. Polymerization of monomer units to form particles
2. Growth of particles
3. Linking of particles into chains, then solid networks that extend throughout the liquid medium, thickening it to a gel

In acid solution, the sol-to-gel transition allows the solid phase to be shaped into films, fibres and monoliths. For preparing coating films and fibres, the sol must exhibit spinnability. It appears that only solutions containing long-chained polymers are spinnable. Films are generally coated on the surface of the substrate via spin coating and dipping processes. Gel fibres are made by fibre drawing from the viscous alkoxide solution at or near room temperature. In contrast, on basic hydrolysis of metal alkoxides, a colloidal sol is generated. The gel is colloidal when the solid network is made of round sol particles.

Removal of the solvents and appropriate drying are important steps to achieve gel densification. When a solvent is evaporated from the gel under atmospheric conditions, capillary pressure due to the interfacial tension of the solvent places a high stress on the gel network. This leads to considerable shrinkage and fracture of the gel during drying. The resultant hard, glassy and porous product is called a *xerogel*. When the liquid within the gel is removed above its critical temperature and pressure (hypercritical) in an autoclave, the capillary pressure can be eliminated. The product thus obtained is referred to as an *aerogel*. An aerogel is generally amorphous and exhibits several unique properties such as high surface areas and porosities, and low densities and conductivities.

The most commonly studied metal alkoxide is silicon tetraethyl orthosilicate [TEOS; $Si(OC_2H_5)_4$]. The TEOS precursor can react readily with water via the following reaction:

$$Si(OR)_4 + H_2O \longrightarrow Si(OR)_3(OH) + ROH$$

As the hydrolysis of silicon alkoxides is very slow, the conversion of metal precursor molecules into trialkoxy silanol, $Si(OR)_3(OH)$ proceeds more rapidly by adding acid or base catalysts. Depending on the amount of water and catalyst present, hydrolysis may proceed partially, in which the metal alkoxides convert into $Si(OR)_{4n}(OH)_n$, or may go to completion in which all OR groups of alkoxides are replaced by OH:

$$Si(OR)_4 + 4H_2O \longrightarrow Si(OH)_4 + 4ROH$$

Subsequent condensation reaction sequences produce cluster species with Si–O–Si (siloxane) bonds and water or alcohol as the by-products:

$$(OR)_3Si - OH + HO - Si(OR)_3 \longrightarrow (OR)_3\,Si - O - Si(OR)_3 + H_2O$$

$$(OR)_3\,Si - OR + HO - Si\,(OR)_3 \longrightarrow (OR)_3\,Si - O - Si(OR)_3 + ROH$$

In silica aerogels, particles of 1–100 nm are arranged in a highly cross-linked silica solid network. In a recent study, colloidal silica particles were synthesized from hydrolysed TEOS, ethanol (C_2H_5OH) and de-ionised water using ammonium hydroxide (NH_4OH) as a catalyst. Silica nanoparticles can also be prepared simply via thermal decomposition, e.g., oxidation of TEOS or rice husk in a reactor, and the CF-CVC route as mentioned previously. Thermal decomposition of rice husk appears to be more cost effective due to global availability of cheap rice husk. Another process involves the precipitation of silica particles from the silicate solution and buffering with sulphuric acid.

Alkoxides of transition metals (e.g., titanium, zirconium) are highly reactive towards water. This implies that the rate of hydrolysis and condensation is very high, and thus it is difficult to form a stable sol of titania (TiO_2). Titania exists in three main crystallographic structures, namely, rutile, anatase and brookite. Among these, rutile is the only stable phase. Many acid catalysts such as nitric acid, hydrochloric acid and acetic acid have been added to lower the reaction rates. Apart from these acids, acetylacetone ($C_5H_8O_2$) can also moderate the reaction rate. It is added as a chelating agent to decrease the reactivity of titanium alkoxides and to form stable colloidal sols of ~ 5 nm. Under more controllable hydrolysis and condensation reactions, smaller particle sizes with more unique properties can be achieved. For example, TiO_2 aerogel nanocrystals can be obtained by supercritical evacuation of solvent from gels prepared through HCl-controlled hydrolysis condensation reactions of titanium isopropoxide in isopropanol. HCl catalyst favours the synthesis of titania powders with smaller grain size compared to acetylacetone catalyst under the same heat treatment conditions. Further, anatase crystallites are more stable if HCl is used as catalyst, while the pure rutile phase can be more easily obtained if acetylacetone is used as a catalyst. Figures 3.7a and 3.7b show the TEM micrographs of nanocrystalline titania prepared by the sol–gel route with a precursor to water ratio of 1 and 4, respectively. Figures 3.7c and 3.7d indicate that the nanocrystallite titania does not significantly grow on calcination up to about 500°C. According to the literature, various metal oxide nanocrystalline powders, e.g., SnO_2, $Ba_2Ti_2O_5$, $PbTiO_3$, (Pb, La)(Zr, Sn, Ti)O_3, nanocomposite powders containing elemental particulates, e.g., nano-Ni/SiO_2, Fe–Al_2O_3 and oxide nanocomposites, e.g., Fe_2O_3–SiO_2, NiO–SiO_2, and $3Al_2O_3$–$2SiO_2$ have been synthesized using the sol–gel process.

The sol–gel process has been useful for synthesizing only metal oxides as a result of the presence of metal–oxygen bonds in the alkoxide precursor, and the resulting gels are essentially metal hydroxides or oxides. This process has distinct advantages over other techniques for preparing metal oxide nanoparticles. These include the formation of high purity powders as a result of homogeneous mixing of the raw materials on the molecular level, and the large-scale industrial production of nanopowders. The disadvantage of the process is the high cost of alkoxide precursors. In some cases, the sol–gel route can also be used to prepare non-oxide ceramic powders such as β-SiC and ferroelectrics like (Pb, La)(Zr, Sn, Ti)O_3 (PLZST). In the

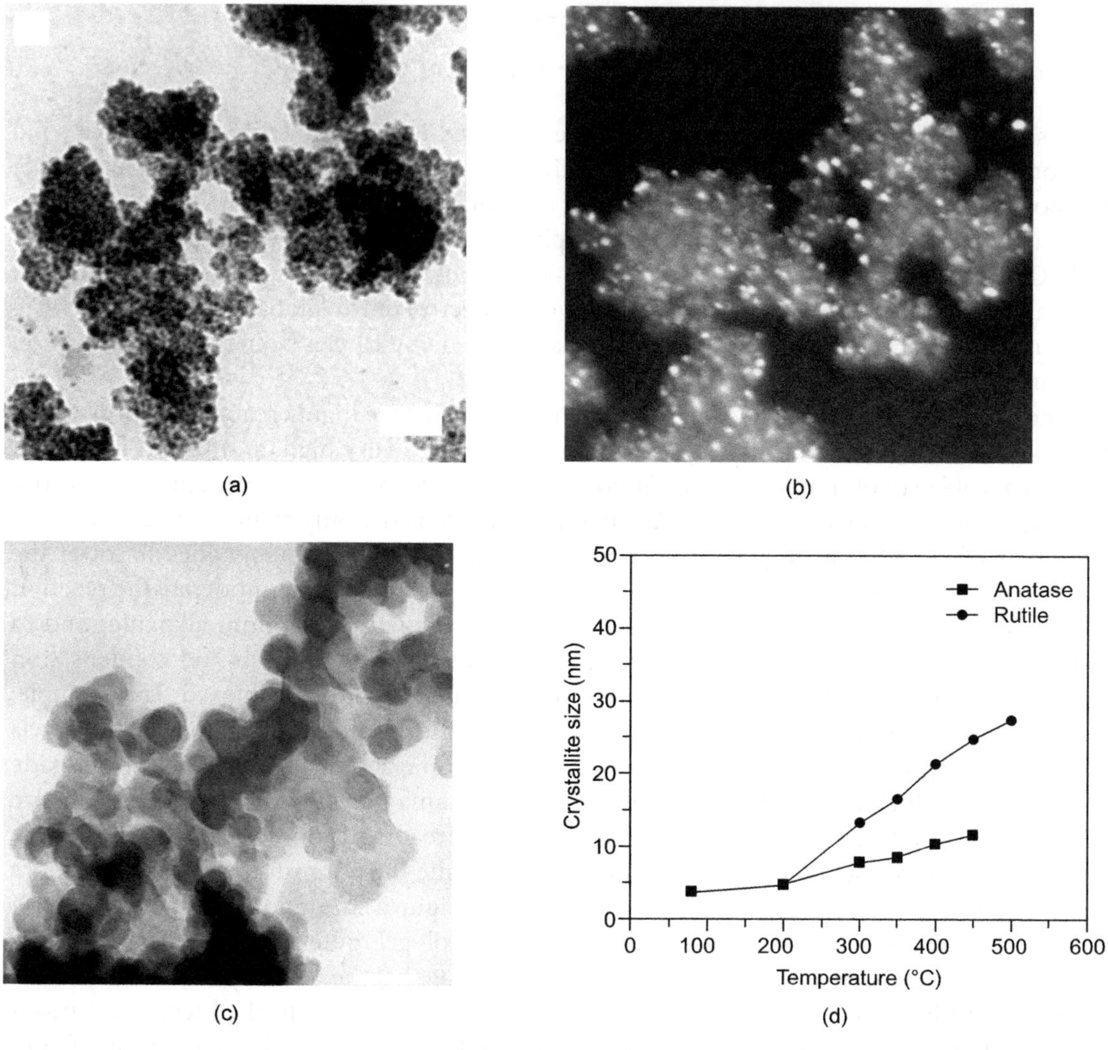

Fig. 3.7 TEM micrographs of nano titania (a) as-prepared (bright field image) with precursor to water ratio of 1, (b) as-prepared (dark field image) with precursor to water ratio of 4, (c) after calcination at 400°C for 2 h (bright field image) with precursor to water ratio of 4 and (d) Variation of the crystallite size of the anatase and rutile phases with calcination temperature with precursor to water ratio of 4. (*Source:* BS Murty, IIT Madras).

former case, heat treatment of the gel product at 1500°C in argon atmosphere resulted in 0.6 mm spherical agglomerates consisting primarily of particles of 40 nm in diameter. Recently, β-SiC nanopowders of 13–30 nm were synthesized by the chemical vapour reaction of the SiH_4–C_2H_4–H_2 system in the temperature range of 1423 K and 1673 K; the products were free from agglomeration. β-SiC powders can be made available at relatively low cost by the process of pyrolising rice husks at 1200–1500°C. The disadvantage is the larger size of β-SiC powders produced by this process, i.e., within micrometre range. It is worth noting that transition metal nanorods can be made by reacting carbon nanotubes with volatile metal oxides or halides. Thus, β-SiC nanorods can be produced from reactions of carbon nanotubes with SiO or Si–I reactants. The preferred growth direction of these nanorods is [111], though rods with [100] growth are also observed at low reaction temperatures.

The anti-ferroelectric ceramics are promising candidates for new ceramic actuators because of their unique characteristics of big and isotropic strain changes under applied electric field, as well as shape memory effects. Nano-sized ceramics prepared by the sol–gel method are usually more homogeneous and reactive than those fabricated by conventional solid-state reactions since the mixing of the reagents occurs on a finer scale. The lower processing temperature is also helpful in minimising the PbO loss. One of the problems in lead-based complex perovskites is the evaporation of PbO at high temperatures, leading to undesirable phases. Sol–gel derived powders of single perovskite phase can be used to produce bulk ceramic samples after sintering at various temperatures.

3.1.5 Wet chemical synthesis

Solution-based processing routes used for the synthesis of nanoparticles include precipitation of solids from a supersaturated solution, homogeneous liquid phase chemical reduction and ultrasonic decomposition of chemical precursors. These processes are attractive due to their simplicity, versatility and availability of low cost precursors. Inorganic salt compounds used in the wet chemical synthesis routes are more versatile and economical than alkoxides employed in the sol–gel process. A typical example is the formation of nanocrystalline titania powders via hydrolysis of $TiCl_4$ at lower temperatures:

$$TiCl_4 + 2H_2O \longrightarrow TiO_2 + 4HCl$$

Once the solution becomes saturated, crystallization of titania takes place either through homogeneous or heterogeneous nucleation. In the latter case, crystal seeds are added to the solutions to promote the crystallization of titania nanoparticles.

Salt reduction is one of the most commonly adopted methods to generate the metal colloid particles. The process involves the dissolution of metal salts in aqueous or non-aqueous environments followed by the reduction of metal cations to the zero-valent state. The nature of the metal salts determines the kind of reducing agent to be applied. To produce transition metal nanoparticles, group 6 metal chlorides such as $CrCl_3$, $MoCl_3$ and WCl_4 are reduced with $NaBEt_3$ in toluene solution at room temperature to form metal colloids of high yield. A typical reaction for forming metal powders is given as:

$$MCl_x + xNaBEt_3 \longrightarrow M + xNaCl + xBEt_3 + \left(\frac{1}{2}x\right) H_2$$

where $x = 3$ for M = Cr and Mo; $x = 4$ for M = W. However, the reduction of the chromium, molybdenum and tungsten halides with either $LiBEt_3H$ or $NaBEt_3$ in tetrahydrofuran (THF) solution generates the corresponding metal carbides (M_2C) rather than the metals. In another experiment, nanostructured $Fe_{50}Ni_{50}$ alloy has been synthesized by ultra-rapid autocatalytic chemical reduction of the corresponding transition metal ions in alkaline aqueous solution with hydrazine hydrate ($N_2H_4 \cdot H_2O$) at 353 K. The alloy powder is composed of spherical particles with 96 nm mean size diameter.

Metal nanoparticles can also be generated via ultrasonic and thermal decomposition of metal salts or chemical precursors. Recently, ultrasonic waves were employed to stimulate the chemical reactions of inorganic salts. Sonication of argon-saturated aqueous solutions of NaAuCl and $PdCl_2$ results in the formation of noble metal AuPd alloy nanoparticles. Power ultrasonic waves can stimulate certain novel chemical processes due to the formation of localised hot spots in the liquid of extremely high temperatures (~ 3000 K) and high pressures (~ 1000 atm). The main event in the process is the nucleation, growth and collapse of cavitation bubbles formed in the liquid. The cooling achieved during the cavitation collapse is estimated to be greater than 2×10^9 K. This process is commonly referred to as the sonochemical method. Transition metal nanoparticles can be produced via sonication of their respective chemical precursors. For example, $Ni(CO)_4$ has been sonicated under argon atmosphere to obtain amorphous nickel. Nanostructured α-Fe is usually synthesized either via sonication or thermal decomposition of $Fe(CO)_5$ solution. One disadvantage of the sonication process is the difficulty in controlling the resulting particle size and distribution due to the agglomeration of particles into a porous coral-like microstructure.

A number of traditional and herbal methods of materials processing could also possibly lead to nanostructure formation. The box overleaf gives the highlights of these routes.

3.1.6 Self-assembly

Self-assembly is a nanofabrication technique that involves aggregation of colloidal nanoparticles into the final desired structure. This aggregation can be either spontaneous (entropic) and due to the thermodynamic minima (energy minimization) energy minimization constraints, or chemical and due to the complementary binding of organic molecules and supramolecules (molecular self-assembly). Molecular self-assembly is one of the most important techniques used in biology for the development of complex functional structures. Since these techniques require that the target structures be thermodynamically stable, they tend to produce structures that are relatively defect-free and self-healing. Self-assembly is by no means limited to molecules or the nano-domain and can be carried out on just about any scale, making it a powerful bottom-up assembly and manufacturing method (multi-scale ordering). Another attractive feature of this technique relates to the possibility of combining the self-assembly properties of organic molecules with the electronic, magnetic and photonic properties of inorganic components.

The central theme behind the self-assembly process is spontaneous (physical) or chemical aggregation of colloidal nano-particles. Spontaneous self-assembly exploits the tendency of mono-dispersed nano- or submicro-colloidal spheres to organise into a face-centred cubic (FCC) lattice. The force driving this process is the desire of the system to achieve a

INDIAN TRADITIONAL METHODS OF NANOSTRUCTURE SYNTHESIS

Science and technology had significantly advanced in India in ancient times. Grant Duff, British Historian of India, observed "Many of the advances in the sciences that we consider today to have been made in Europe were in fact made in India centuries ago." One such Indian heritage in materials science is the rustless wonder, the Delhi Iron pillar, which has been standing for 1600 years without getting corroded. Another example of Indian excellence is Wootz steel, which is an extremely strong steel developed in India around 300 BC. This steel has been extensively used for making swords as it could cut a metal helmet into two pieces (popularly known as Damascus steel). Recently (*Nature*, 2006) Paufler's group from Technical University Dresden, Germany, reported the presence of carbon nanotubes and nanowires in Damascus steel. Similarly, it is also possible to produce carbon nanotubes and nanoparticles from

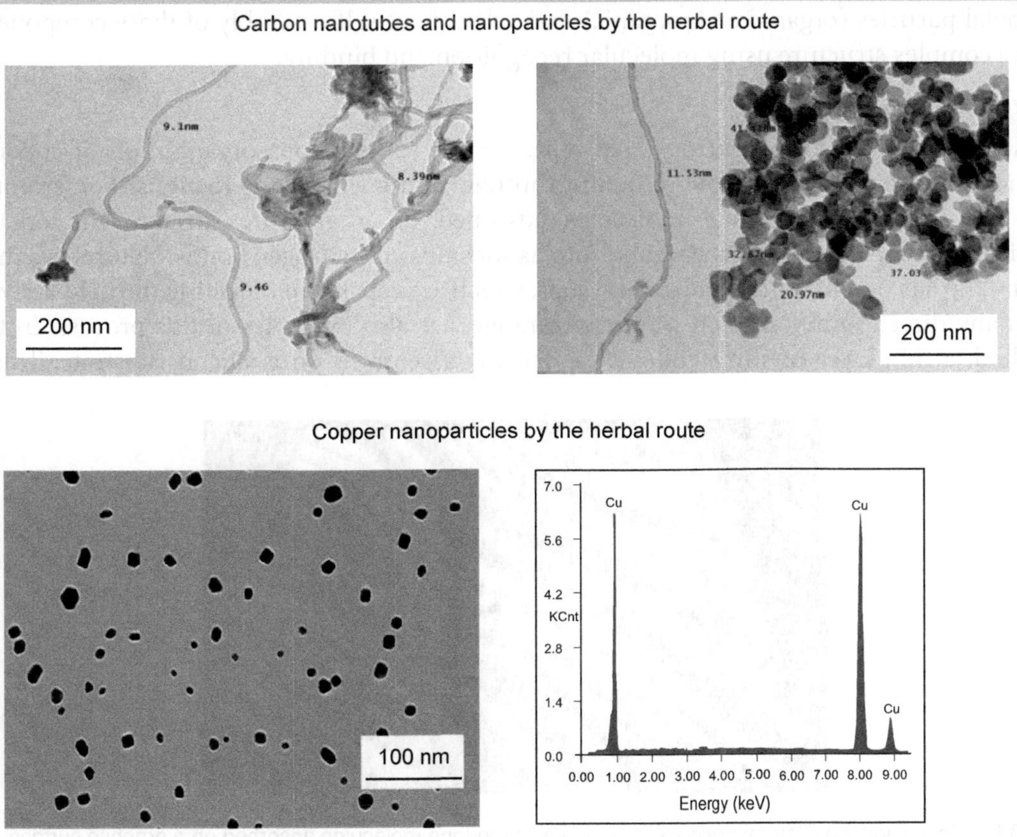

(*Source:* BS Murty, IIT Madras)

various vegetable oils, as shown in the images below. Such nanostructures appear to have been traditionally prepared in India.

Recently a Sanskrit scholar, Dr ABS Sastry, demonstrated the formation of copper nanoparticles by a simple chemical displacement reaction, where traditional herbs have been successfully used to cap the nanoparticles and prevent them from agglomeration. The transmission electron bright field image confirms the nanocrystalline nature of these copper particles. The energy dispersive x-ray spectrum from these individual nanoparticles confirms that they are pure copper.

thermodynamically stable state (minimum free energy). In addition to spontaneous thermal self-assembly, gravitational, convective and electro-hydrodynamic forces can also be used to induce aggregation into complex 3D structures. Chemical self-assembly requires the attachment of a single molecular organic layer (*self-assembled monolayer*, or *SAM*) to the colloidal particles (organic or inorganic) and subsequent self-assembly of these components into a complex structure using molecular recognition and binding.

Physical self-assembly

This method is driven by entropy that relies on the spontaneous organisation of colloidal particles into a relatively stable structure through non-covalent interactions. For example, Fig. 3.8 shows quinacridone molecules adsorbed on a graphite surface. The organic semiconductor molecules self-assemble into nanochains via hydrogen bonds. Natural assembly of on-chip silicon photonic band gap crystals, which are capable of reflecting the light arriving from any direction in a certain wavelength range, are also examples of this process. In this method, a thin layer of silica colloidal spheres is assembled on a silicon substrate. This is

Fig. 3.8 Scanning tunnelling microscope image of quinacridone molecules adsorbed on a graphite surface. The organic semiconductor molecules self-assembled into nanochains via hydrogen bonds. (*Source:* http://commons.wikimedia.org/wiki/*File:*Selfassembly_Organic_Semiconductor_Trixler_LMU.jpg.).

achieved by placing a silicon wafer vertically in a vial containing an ethanolic suspension of silica spheres. A temperature gradient across the vial aids the flow of silica spheres. Once such a template is prepared, LPCVD can be used to fill the interstitial spaces with Si, so that the high refractive index of silicon provides the necessary band gap.

One can also deposit colloidal particles in a patterned substrate (template-assisted self-assembly, TASA). This method is based on the principle that when an aqueous dispersion of colloidal particles is allowed to de-wet from a solid surface that is already patterned, the colloidal particles are trapped by the recessed regions and assemble into aggregates of shapes and sizes determined by the geometric confinement provided by the template. The patterned arrays of templates can be fabricated using conventional contact mode photolithography, which gives control over the shape and dimensions of the templates, thereby allowing the assembly of complex structures from colloidal particles. This method can be used to fabricate a variety of polygonal and polyhedral aggregates that are difficult to generate.

Chemical self-assembly

Organic and supramolecular (SAM) molecules play a critical role in colloidal particle self-assembly. SAMs are robust organic molecules that are chemically adsorbed onto solid substrates. Most often, they have a hydrophilic (polar) head that can be bonded to various solid surfaces and a long, hydrophobic (non-polar) tail that extends outward. SAMs are formed by the immersion of a substrate in a dilute solution of the molecule in an organic solvent. The resulting film is a dense organization of molecules arranged to expose the end group. The durability of a SAM is highly dependent on the effectiveness of the anchoring to the surface of the substrate. SAMs have been widely studied, because the end group can be functionalised to form precisely arranged molecular arrays for various applications ranging from simple, ultra-thin insulators and lubricants to complex biological sensors. Chemical self-assembly uses organic or supramolecular SAMs as the binding and recognition sites for fabricating complex 3D structures from colloidal nanoparticles. Most commonly used organic monolayers include:

- Organosilicon compounds on glass and native surface oxide layer of silicon
- Alkanethiols, dialkyl disulfides and dialkyl sulfides on gold
- Fatty acids on alumina and other metal oxides
- DNA

Octadecyltrichlorosilane (OTS) is the most common organosilane used in the formation of SAMs, mainly because it is simple, readily available and forms good, dense layers. Alkyltrichlorosilane monolayers can be prepared on clean silicon wafers with a surface of SiO_2 (with almost 5×10^{14} SiOH groups/cm^2). Since the silicon chloride bond is susceptible to hydrolysis, a limited amount of water has to be present in the system in order to obtain good quality monolayers. Monolayers made of methyl- and vinyl-terminated alkylsilanes are autophobic to the hydrocarbon solution and hence emerge uniformly dry from the solution, whereas monolayers made of ester-terminated alkylsilanes emerge wet from the solution used in their formation. The disadvantage of this method is that if the alkyltrichlorosilane in the solvent adhering to the substrate is exposed to water, a cloudy film is deposited on the surface due to the formation of a gel of polymeric siloxane.

Another important organic SAM system is the alkanethiols $X(CH_2)_nSH$, where X is the end group) on gold. A major advantage of using gold as the substrate material is that it does not have a stable oxide, and can thus be handled in ambient conditions. When a fresh, clean, hydrophilic gold substrate is immersed (several minutes to several hours) in a dilute solution (10^{-3} M) of the organic sulphur compound (alkanethiols) in an inorganic solvent, a close-packed, oriented monolayer can be obtained. Sulphur is used as the head group, because of its strong interaction with gold substrate (44 kcal/mol), resulting in the formation of a close-packed, ordered monolayer. The end group of alkanethiol can be modified to render hydrophobic or hydrophilic properties to the adsorbed layer.

Another method for depositing alkanethiol SAM is soft lithography. This technique is based on inking a PDMS stamp with alkanethiol and its subsequent transfer to planar and non-planar substrates. Alkanethiol functionalised surfaces (planar, non-planar, spherical) can also be used to self-assemble a variety of intricate 3D structures.

Carboxylic acid derivatives self-assemble on surfaces (e.g., glass, Al_2O_3 and Ag_2O) through an acid–base reaction, giving rise to monolayers of fatty acids. The time required for the formation of a complete monolayer increases with decreased concentration. A higher concentration of carboxylic acid is required to form a monolayer on gold compared to Al_2O_3. This is due to differences in the affinity of the COOH groups (more affinity to Al_2O_3 and glass than gold) and also the surface concentration of the salt forming oxides in the two substrates. In the case of amorphous metal oxide surfaces, the chemisorption of alkanoic acids is not unique. For example, on Ag_2O the oxygen atoms bind to the substrate in a nearly symmetrical manner, resulting in ordered monolayers with a chain tilt angle from the surface normally of 15 to 25°. But on CuO and Al_2O_3, the oxygen atoms bind themselves symmetrically and the chain tilt angle is close to 0°. The structure of the monolayers is thus a balance of the various interactions taking place in the polymer chains.

Deoxyribonucleic acid (DNA), the framework on which all life is built, can be used to self-assemble nanomaterials into useful macroscopic aggregates that display a number of desired physical properties. DNA has a double-helix structure with two strands coiled around each other. When the two strands are uncoiled, singular strands of nucleotides are left. These nucleotides consist of a sugar (pentose ring), a phosphate (PO_4) and a nitrogenous base. The order and architecture of these components are essential for the proper structure of a nucleotide.

There are typically four nucleotides found in DNA: adenine (A), guanine (G), cytosine (C) and thymine (T). A key property of the DNA structure is that the described nucleotides bind specifically to another nucleotide when arranged in the two-strand double helix (A to T and C to G). This specific bonding capability can be used to assemble nanophase materials and nanostructures. For example, nucleotide functionalised nano-gold particles have been assembled into complex 3D structures by attaching DNA strands to the gold via an enabler or linker. In a separate work, DNA was used to assemble nanoparticles into macroscopic materials. This method uses alkane dithiol as the linker molecule to connect the DNA template to the nanoparticle. The thiol groups covalently attach themselves to the colloidal particles, leading to aggregate structures.

3.2 TOP-DOWN APPROACHES

3.2.1 Mechanical alloying

The mechanical alloying (MA)/milling process was originally developed by Benjamin of the International Nickel Company for the production of oxide dispersion strengthened (ODS) superalloys. It is now a widely used process for the fabrication of nanocrystalline powders. The possible spectrum of the uses of MA are shown schematically in Fig. 3.9. Recently, nanocrystalline high-entropy solid solutions with high hardness have been synthesized in multicomponent equiatomic alloys by MA.

Mechanical alloying or milling is usually carried out in high-energy mills such as vibratory mills (Spex 8000 mixer/mill), planetary mills (Fritsch and Retsch) and attritor mills (Szegvari attritor and Simoloyer). The vibratory mill has one vial, containing the sample and grinding balls and vibrates in all three directions. Because of the amplitude (about 50 mm) and speed (about 1200 rpm), the ball velocities are high (in the order of 5 m/s) and consequently the force of the ball's impact is unusually high. Another popular mill for conducting MA experiments is the planetary ball mill. In this mill, the vials rotate around their own axes and at the same time around the axis of a disc on which they are mounted. The centrifugal force produced by the vials rotating around their own axes and that produced by the rotating support disk together act on the vial contents, consisting of material to be ground and the grinding balls. Since the vials and the supporting disc rotate in opposite directions, the centrifugal forces alternately act in opposite directions. Due to this, the balls run down the inside wall of the vial. This is followed by the material being ground and grinding balls being lifted off and travelling freely through the inner chamber of the vial and colliding against the opposing inside wall. Grinding vials and balls are available in eight different materials—agate, silicon nitride, sintered corundum, zirconia, chrome steel, Cr–Ni steel, tungsten carbide and plastic polyamide.

An attritor ball mill consists of a stationary vertical drum in which a vertical shaft rotates with a series of horizontal impellers attached to it. Set progressively at right angles to each other, the impellers, through their rotation, energise the ball charge, causing powder size reduction

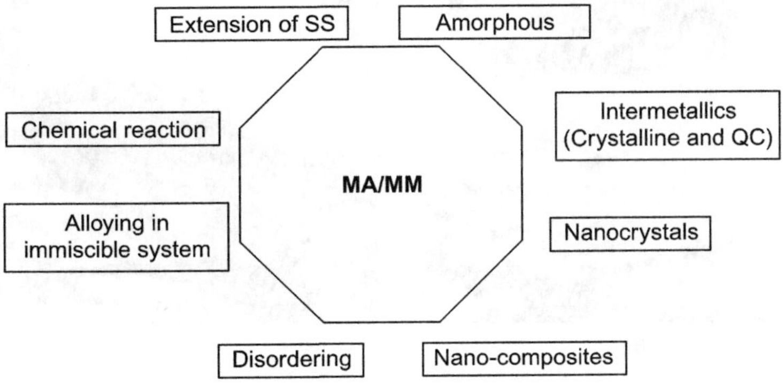

Fig. 3.9 Attributes of mechanical alloying or high-energy ball milling.

due to the impact between balls, between balls and the container wall, and between balls, the agitator shaft and impellers. Particle size reduction also occurs partially by interparticle collisions and by ball sliding on the walls of the vials. Attritors are the mills in which large quantities of powder (from about 0.5 to 40 kg) can be milled at a time. The most recent of the ball mills is the horizontal attritor (Simoloyer) that can be operated in dry processing at high relative velocity of the grinding media (up to 14 m/s) under controlled condition like vacuum or inert gas and in closed circuits. In these mills, the grinding media is accelerated by a horizontally arranged rotor inside the grinding vessel. These mills have the advantage of highest relative velocity of grinding media, which leads to high level of kinetic energy transfer, an intensive grinding effect and short processing times. The short processing times and collision-based grinding process results in low contamination levels. The simoloyers are available with 0.5- to 990-litre grinding chamber capacity, which makes it very convenient to scale-up the laboratory experiments to commercial production plants.

The mechanism of nanocrystallization during high-energy ball milling was first proposed by H.J. Fecht in 1983. He summarised the phenomenon of the development of nanocrystalline microstructure into three stages;

Stage 1: Localization of deformation into shear bands with high dislocation density.

Stage 2: Dislocation, annihilation and recombination to form nanometre-scale subgrains, which extend throughout the sample with further milling.

Stage 3: Transformation of subgrain boundary structure to randomly oriented high-angle grain boundaries. Superplastic deformation processes such as grain boundary sliding causes self-organisation into a random nanocrystalline state.

During high-energy milling, the powder particles are repeatedly flattened, cold welded, fractured and re-welded. Whenever balls collide, some amount of powder is trapped in between them. The impact from the balls causes plastic deformation of the powder particles, causing work hardening and fracture. The new surfaces formed by the fracture of the particles weld together. The composite particles at this stage have a characteristic layered structure consisting

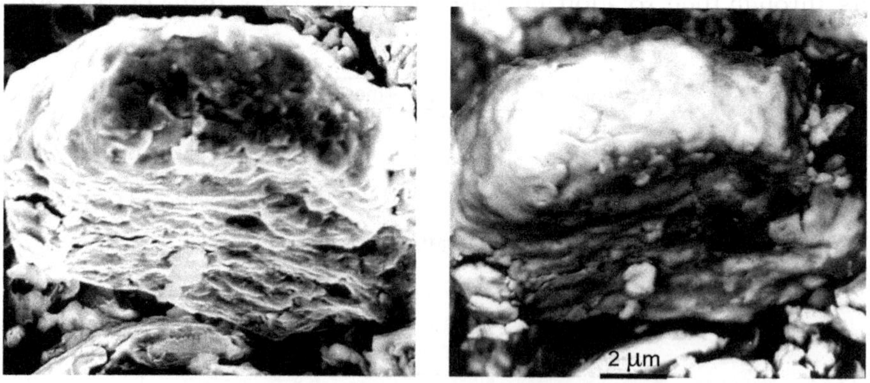

(a) (b)

Fig. 3.10 (a) Secondary electron image and (b) backscattered electron image of layered structure observed during initial stages of high-energy ball milling of Ti and Al powder particles (*Source:* BS Murty, IIT Madras).

of various combinations of the starting constituents and are shown in Figs. 3.10a and 3.10b for the Ti–Al system. As deformation continues, the particles become work hardened and fracture. When fracture predominates over cold welding, the particle size is refined.

The minimum grain size achievable by high-energy ball milling has been related to several physical properties of elemental metals. It is reported that the minimum grain size, d_{min}, is inversely proportional to the melting temperature for low melting FCC metals. However, for HCP and BCC systems as well as FCC elements with melting temperature above that of Pd (1555°C), d_{min} appears to be insensitive to the melting point. For FCC metals, an inverse relationship was also found between d_{min} and bulk modulus, and a direct relationship between d_{min} and critical equilibrium distance between two edge dislocations. The majority of nanocryztalline metals have been synthesised to understand the mechanism of nanocrystallization and minimum grain size obtained by high-energy ball milling. Research has also been focussed on the validity of the Hall–Petch relation, and mechanical and physical properties of nanocrystalline metals. Fe, Cu, Ni, Pd, Cr, Nb, W, Hf, Ru, Zr, Co, Ag, Al, Si and graphite are among the few mentioned. Fecht et al, presented the first systematic report on the synthesis of nanocrystalline metals (Fe, Cr, Nb, W, Zr, Hf, Co and Ru) by high-energy ball milling. It is postulated that the grain boundary energy of the ball-milled nanocrystalline materials is higher than that of fully equilibrated grain boundaries.

An allotropic and polymorphic transformation induced by high-energy ball milling has been widely studied. High-energy ball milling of elemental Nb to a crystallite size of below 10 nm resulted in an allotropic transformation of BCC-Nb to FCC-Nb. Similar allotropic transformations such as HCP-Zr to FCC-Zr and HCP-Ti to FCC-Ti have been observed during high-energy ball milling by Manna and his group at IIT, Kharagpur. FCC to HCP transformation in Ni during high-energy ball milling was found to be aided by the addition of small amounts of Si and Al, which when dissolved in Ni can make it harder. A harder material can become nanocrystalline much more easily during high-energy ball milling due to extensive fracture when compared to a ductile material. Thus, the critical grain size of about 10 nm for the FCC to HCP transition in Ni could be achieved early during milling. It has been argued that the structural instability due to plastic strain, increasing lattice expansion and negative hydrostatic pressure was responsible for such transformations. Polymorphic transformations were also observed during high-energy ball milling of the $Ti_{50}Zr_{50}$ binary alloy. Transformation of quasicrystalline phase to crystalline BCC phase has also been observed in the Al–Cu–Fe and Al–Cu–Co systems. It was even observed in a number of systems such as Ti–Al, Ti–Ni, Ti–Cu, Fe—Si, etc., that nanocrystalline alloys below a certain critical size of about 10 nm (and above a critical lattice strain) can get structurally destabilised and become amorphous due to the large defect concentration.

Nanocrystallization by high-energy ball milling has also been demonstrated to help in extending the solid solubility of different solute elements into various solvent elements, as discussed in the previous chapter. This is basically due to the large fraction of grain boundaries in these nanocrystalline materials, which can accommodate large solute fractions.

Many ordered intermetallic compounds show high yield strength up to high temperatures and have the advantage of lower density compared to commercial superalloys. However, these advantages are hampered due to their brittleness at low temperatures. It has been shown that the nanocrystalline intermetallics have improved formability. This finding has created an interest in the synthesis of several nanocrystalline aluminides, silicides and other intermetallics through MA. Interestingly, it has been observed that the phase fields of intermetallics can be significantly enhanced in the nanocrystalline state. In case of nanocrystalline NiAl synthesized by MA, the phase field NiAl has been extended from 45–55 at.%Ni in the conventional microcrystalline state to 25–65 at.%Ni in the nanocrystalline state. Similarly, studies on the Ni–Si, Fe–Si, Ni–Al, Al–Cu systems have indicated that only congruent melting compounds (which are expected to have lower surface energies) are expected to be stable in the nanocrystalline state.

It has also been observed that polymorphic transitions can be induced in intermetallics in the nanocrystalline state. Nayak and Murty synthesized the nanocrystalline $L1_2$–Al_3Ti intermetallic compound with the crystallite size less than 15 nm by MA of $Al_{67}M_8Ti_{25}$ (M = Cr, Mn, Fe, Co, Ni and Cu), while Al_3Ti usually has a DO_{23} structure. Among all the elements studied, Cu appears to be the best to stabilize $L1_2$–Al_3Ti. Another interesting observation was that the $L1_2$–Al_3Ti phase formed at an Al crystallite size of 15 nm. Similarly, $L1_2$–Al_3(Ti, Zr) compounds have formed at an Al crystallite size of about 20 nm, irrespective of the composition, as shown in Fig. 3.11. These observations clearly indicate that nanocrystallization is a prerequisite for the formation of intermetallic compounds. Figures 3.12a and 3.12b show the TEM dark field image and corresponding SAD pattern, respectively, obtained from nanocrystalline Al_3Zr.

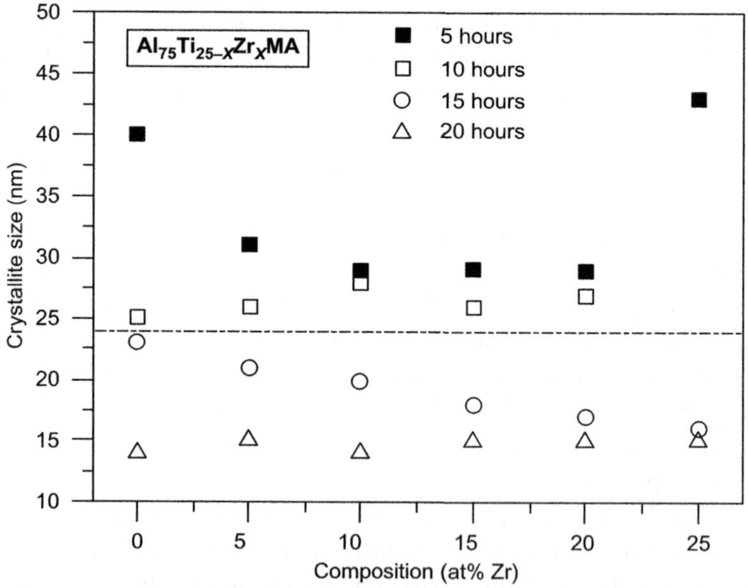

Fig. 3.11 Crystallite size of Al and intermetallic as a function of composition. Nanocrystalline intermetallic compound formation was observed when the crystallite size of Al was below 24 nm. The dotted line demarcates the phase field of mixture of elements and intermetallic compound. (*Source:* BS Murty, IIT Madras).

In situ nanocomposite formation by MA has found significant interest due to their potential application. It has been shown by Naser and his co-workers that no grain growth occurs in the matrix close to its melting point when Cu and Mg are reinforced with nanocrystalline Al_2O_3 by MA. In another interesting study, nanocomposites in the Al–Fe system have been synthesized by MA, which have shown exceptionally high hardness of about 13 GPa with densities very close to theoretical density. The very high hardness in these nanocomposites has been attributed to the presence of nanocrystalline/quasicrystalline intermetallic phases. Figure 3.13a shows the bright field image of nanocomposites in the Al–20Fe alloy. Detailed

Fig. 3.12 TEM dark field image and corresponding diffraction pattern of $Li1_2$–Al_3Ti intermetallic. (*Source:* BS Murty, IIT Madras).

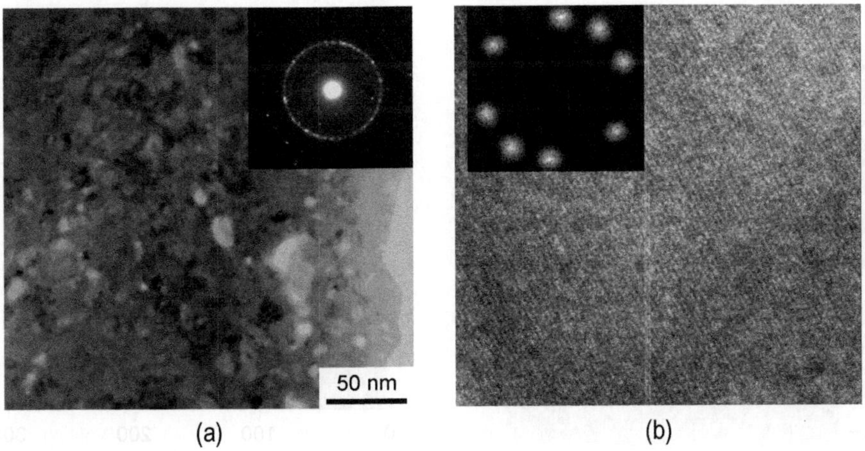

Fig. 3.13 (a) TEM bright field image of images of nanocomposites in Al–20Fe alloy after MA and subsequent annealing at 673 K for 2 h and (b) high resolution image, showing lattice fringes from nano-quasicrystalline particles in the Al matrix. (*Source:* BS Murty, IIT Madras).

high-resolution electron microscopy of these nanocrystalline intermetallics in this alloy indicated the possible presence of nano-quasicrystalline phase in these alloys, as shown in Fig. 3.13b.

Cu is a good electrical conductor; however, it has low strength due to which the life of Cu conductors as electrodes and as electrical contacts is quite short. Increasing the strength of Cu by using it in nanocrystalline form can improve the life of these conductors. However, nanocrystalline pure Cu can have significant grain growth at high temperatures. Cu-based nanocomposites such as Cu–W and Cu–Ta would be the answer to this problem. Figure 3.14a shows TEM evidence of the nanocrystalline nature of these Cu–20wt.%Ta nanocomposites. The AFM analysis of nanocomposites after sintering clearly showed that W and Ta can restrict the grain growth in Cu and retain the nanocrystalline grains. Figure 3.14b represents a typical AFM image of Cu–20wt.%W after sintering at 500°C. This indicates that even after sintering at about half the melting point of Cu, the nanocrystalline nature is retained in these nanocomposites. Interestingly, the nanocomposites could be sintered close to their theoretical densities at 500°C, while the micro-composies had to be sintered at temperatures close to the melting point of Cu for achieving similar densities as shown in Figs. 3.15a and 3.15b. These metal–metal nanocomposites synthesized by high-energy ball milling showed superior hardness and compressive strength without much reduction in conductivity. The Cu–W nanocomposites showed ~ 2.6 times higher hardness than microcomposites of similar composition. A similar observation is also made with Cu–Ta nanocomposites. Figure 3.16a shows the compressive strength of the Cu–Ta nanocomposites and microcomposites. The figure also shows the strength contribution (according to Hall–Petch equation) of the nanocrystalline grains. The Cu–Ta nanocomposites exhibited 3.6 to 5 times higher compressive strength than the microcrystalline composites of similar composition. The compressive strength of Cu–Ta nanocomposite is 4.5 to 7 times higher

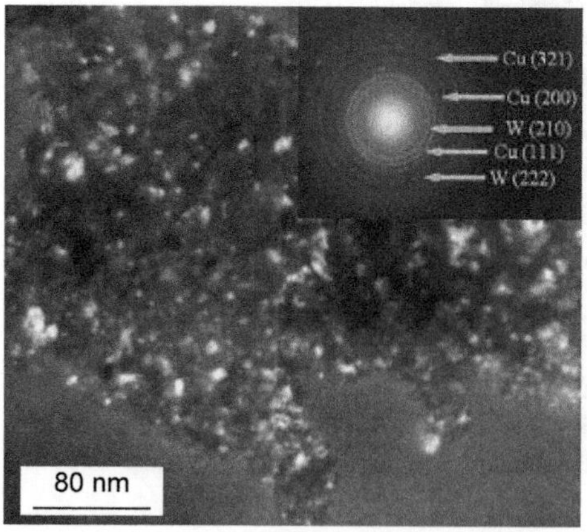

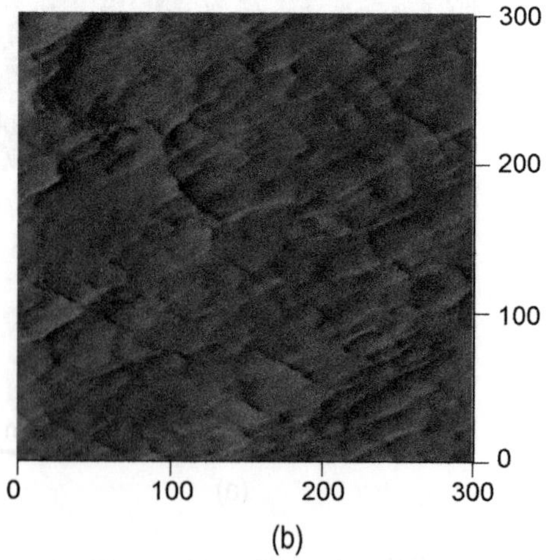

Fig. 3.14 (a) TEM image of Cu-20wt.%W ball-milled powders after 20 h of ball milling; (b) AFM analysis revealed nanograins in Cu-20wt.%W nanocomposites after sintering at 500°C for 2 h. (*Source:* BS Murty, IIT Madras).

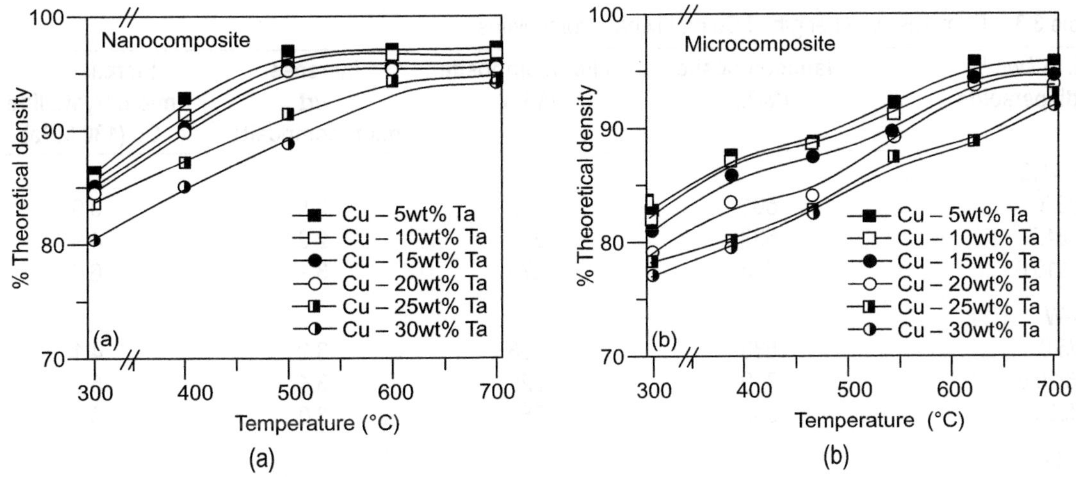

Fig. 3.15 Variation of density of Cu–Ta (a) nanocomposites and (b) microcomposites as a function of sintering temperature. (*Source:* BS Murty, IIT Madras).

than that of microcrystalline Cu, as shown in Table 3.3. The resistivity values of Cu–Ta nanocomposites (Fig. 3.16b) indicate that the major contribution of resistivity is due to Ta rather than fine grain size. Figure 3.17 shows the hardness to resistivity ratio of Cu–Ta nanocomposites normalized with that of the oxygen-free high conductivity (OFHC) Cu, which clearly indicates that the hardness on nanocomposites is three times higher than that of OFHC Cu for similar resistivity.

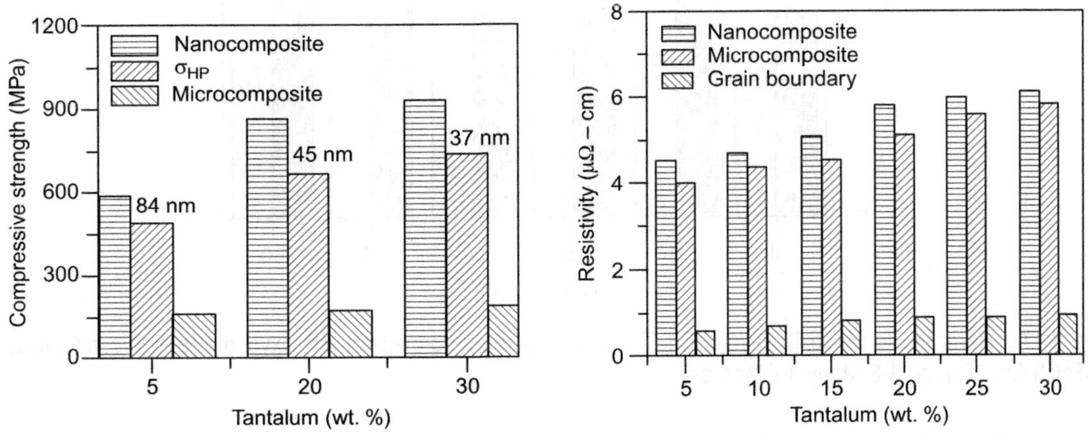

Fig. 3.16 (a) Compressive strength and (b) electrical resistivity of Cu–Ta nanocomposites synthesized by high-energy ball milling. (*Source:* BS Murty, IIT Madras).

Table 3.3 Compressive strength of Cu matrix nanocomposites

Vol. % / wt. % of dispersoids	Nanocomposite (MPa)	Microcomposite (MPa)	Increase wrt microcomposite	Increase wrt microcrystalline Cu (130 MPa)
Cu–Al_2O_3				
5.0/2.3	501	182	2.8	3.9
10.0/4.7	633	221	2.9	4.9
20.0/10.0	853	275	3.1	6.6
Cu–W				
2.4/5.0	536	168	3.2	4.1
11.6/20.0	770	229	3.4	5.9
16.5/30.0	853	284	3.0	6.6
Cu–Ta				
2.7/5.0	586	164	3.6	4.5
11.8/20.0	870	174	5.0	6.7
18.7/30.0	930	185	5.0	7.1

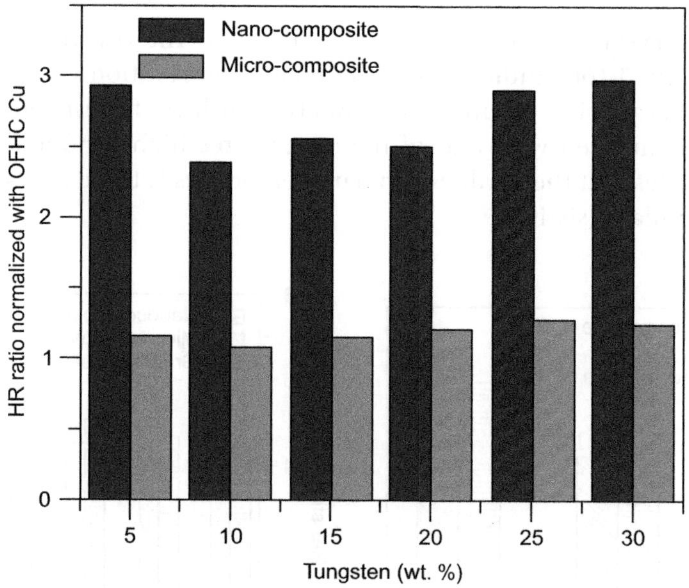

Fig. 3.17 Hardness to resistivity ratio of Cu–Ta nanocomposites normalized with oxygen-free high conductivity (OFHC) Cu. (*Source:* BS Murty, IIT Madras).

3.2.2 Equal channel angular pressing (ECAP)

The principle of ECAP [or equal channel angular extrusion (ECAE)] is illustrated schematically in Fig. 3.18. For the die shown in the illustration, the internal channel is bent through an abrupt angle, ϕ, equal to 90°, and there is an additional angle, ψ, equal to 0° in Fig. 3.19, which represents the outer arc of curvature where the two channels intersect. In this process, a rod-like sample is pressed through the die using a plunger. Shear deformation occurs as the sample passes through the die. Despite the introduction of an intense strain as the sample passes through the shear plane, the sample ultimately emerges from the die without experiencing any change in the cross-sectional dimensions.

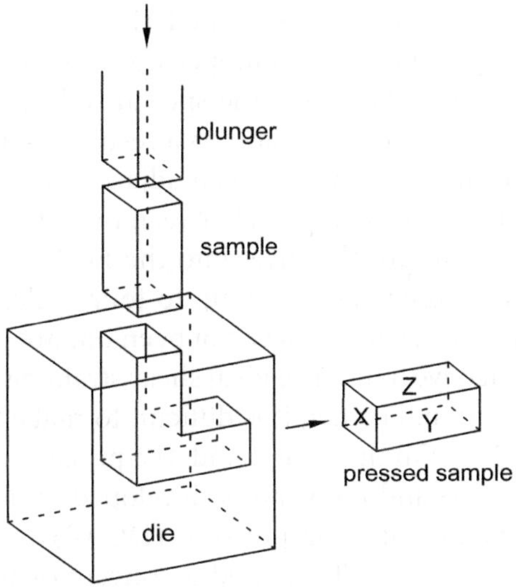

Fig. 3.18 Schematic illustration of ECAP facility.

Researchers all over the world are now studying the ECAE process as it is easy to install and does not require special equipment in the laboratories, except for the dies and a press machine. The ECAE process has been applied to various kinds of metals and alloys and has succeeded in producing ultrafine grain (UFG) microstructures (Fig. 3.19). However, the major materials used are light metals like Al- and Mg-alloys and the trials for steels are limited. This is presumably because large force is required to put the materials through the channel die, overcoming large flow stress and frictional stress. The hydrostatic compressive stress field in the ECAE process is an advantageous point to prevent fracture of the materials. However, cracking due to shear localization has been reported in less workable materials such as Ti–Al–V alloy.

Though the ECAE can fabricate bulk materials, the typical size of the samples is small. Further, the ECAE is principally not a continuous process but a batch process. A few experiments to make the ECAE continuous have been attempted (Fig. 3.20). Saito and

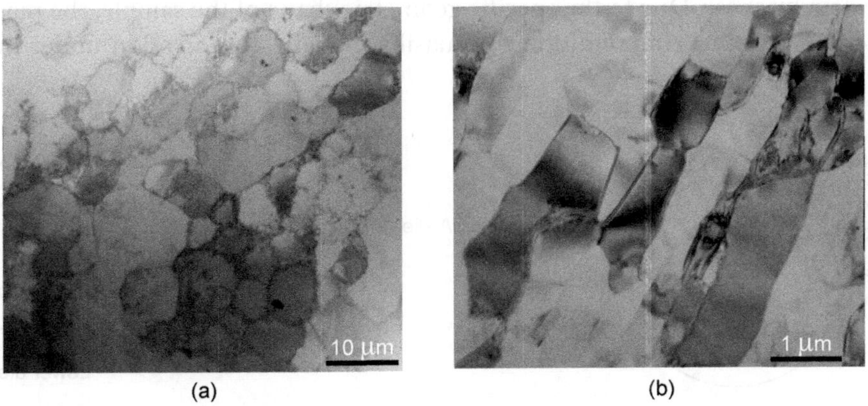

Fig. 3.19 Microstructure of (a) annealed CP aluminum and (b) after ECAP. (*Source:* Uday Chekkingal, IIT Madras).

his co-workers developed the *conshearing process* for continuous ECAE of sheet metals (Fig. 3.20a). They equipped the ECAE die at the end of the satellite mill they previously invented. The *satellite mill* is the special rolling mill which can maintain the rotating speed of all the satellite rolls constant. As a result, compression force along the rolling direction (RD) appears in the materials between adjacent satellite rolls. Folding of the materials is prevented by the guide shoe equipped between the satellite rolls. The compression force in the satellite mill was used to put the sheet into the ECAE die. The conshearing was applied to obtain aluminium of commercial purity up to four cycles and succeeded in fabricating sound sheets, but UFG microstructures have not been obtained. Because they used sheet material, it was probably not effective to impose ideal shear strain owing to the bending and bending-back deformation and friction. Further, it is difficult to make the channel angle 90° in this configuration (they used 125°). On the other hand, the processed materials showed unique textures.

Lee and co-workers developed another continuous ECAE process, named the dissimilar channel angular pressing (DCAP) process, which is principally the same as conshearing (Fig. 3.20b). They used a conventional two-high mill, but the surface of the lower roll was mechanically roughened in order to feed the material into the ECAE die. As a result, the surface quality of the specimens would be worse than the conventionally rolled materials. Strips of 1.55 × 20 × 1000 mm were processed by the continuous confined strip shearing process and the UFGs similar to those obtained in the conventional ECAE have been reported in 1050-Al, though the difficulties and the disadvantages are not obvious from the limited publications currently available. The channel angle was varied from 100° to 140°, and it was found that the critical strain to form the UFGs increases by increasing the angle.

3.2.3 High-pressure torsion (HPT)

A method for fabrication of disc-type bulk nanostructured samples using high-pressure torsion is shown in Fig. 3.21. In HPT, the sample disk is strained in torsion under the applied pressure (*P*) of several GPa between two anvils. A lower holder rotates and surface friction forces the ingot to deform by shear. Due to the specific geometric shape of the sample, the main volume of the material is strained in conditions of quasihydrostatic compression under the applied

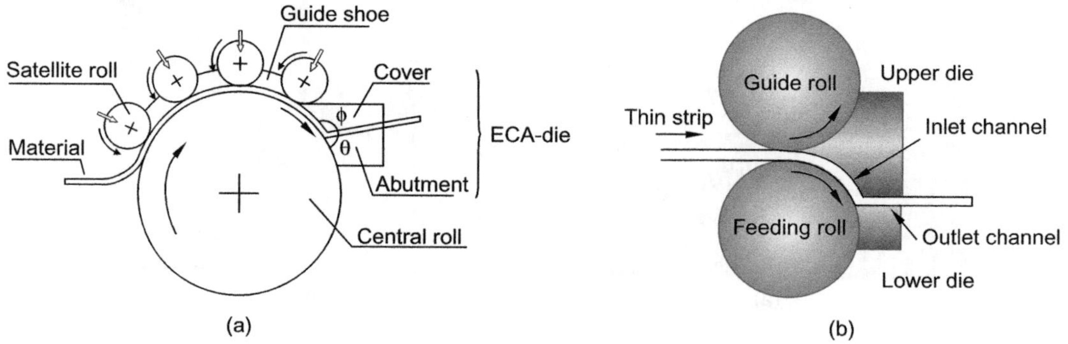

Fig. 3.20 Principle of (a) conshearing and (b) dissimilar channel angular pressing process.

pressure and the pressure of sample outer layers. As a result, in spite of large strain values, the deformed sample is not destroyed. Various strains can be achieved by rotating the lower anvil through a defined angle. The shear strain γ may be simply estimated as,

$$\gamma = \frac{2\pi RN}{d}$$

where R is the distance from the sample centre, N is the number of anvil rotations and d is the thickness of the sample. In this method of microstructure refinement, the simple shear stress conditions are realised at relatively low temperatures under high applied pressure (5–15 GPa). The samples fabricated by severe torsion straining are usually disc-shaped, with a diameter of 10 to 20 mm and thickness 0.2±0.5 mm. A significant change in the microstructure is observed already after deformation by 1/2 rotation, but for formation of the homogeneous nanostructure, as a rule, several rotations are required.

Using high-pressure torsion, the first nanocrystalline structures were developed in single crystals of Ni and Cu. It was found that the mean grain size had gradually decreased with increasing shear strain and had finally stabilized at the level of ~ 100 nm. Further experiments with initially coarse grained Cu, Ni, Fe, Cr and Ti revealed that strong grain refinement can be achieved in all these metals after 3–5 revolutions, whereupon the mean grain size usually reaches the steady state value of 100–200 nm, depending on the material (Table 3.4). Slightly smaller grain size has been reported for single phase matrix Fe and Al alloys and significantly smaller grain size, down to 10 nm, can be obtained for some multiphase alloys and intermetallic compounds.

Recent investigations also showed that severe torsion straining can be used successfully not only for the refinement of a microstructure but also for the consolidation of powders. It was revealed that during torsion straining at room temperature, high pressures equal to several GPa can provide a rather high density close to 100% in the processed disc-type nanostructured samples. For fabrication of such samples via severe torsion straining

Table 3.4 Grain size of various kinds of HPT processed metals

Metal	Details of HPT processing	Grain size (nm)
Ni	$P = 6$ GPA, $\gamma = 300$, RT	170
Cu	$P = 5$ GPA, $\gamma = 300$, RT	200
Cu	$P = 8$ GPA, $\gamma = 250$, RT	200
Fe	$P = 5$ GPA, $\gamma = 600$, RT	100
Fe	$P = 5$ GPA, $\gamma = 420$, RT	127
Cr	$P = 7.8$ GPA, $\gamma = 120$, $T = 623$ K	500
Mo	$P = 6$ GPA, $\gamma = 300$, $T = 623$ K	190
Ti	$P = 5$ GPA, $\gamma = 600$, RT	120

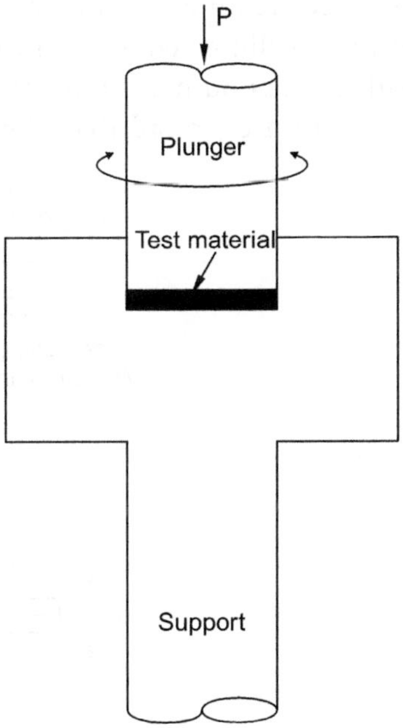

Fig. 3.21 Principle of high pressure torsion.

consolidation, not only conventional powders but also powders subjected to ball milling can be used. The HPT consolidation of nanostructured Ni powder prepared by ball milling can be given as an example. The conducted investigations showed that the density of the fabricated powders is close to 95% of the theoretical density of the bulk coarse-grained Ni. TEM examinations showed the absence of porosity with a mean grain size of about 20 nm. It is also very interesting that the value of microhardness of the Ni samples fabricated by HPT consolidation is as high as 8.6 GPa.

3.2.4 Accumulative roll bonding (ARB)

Accumulative roll bonding is the only SPD process using rolling deformation itself (Fig. 3.22). Rolling is the most advantageous metal working process for continuous production of plates, sheets and bars. However, the total reduction applied to the materials is substantially limited because of the decrease in the cross-sectional dimension of the materials with increasing reduction. In the ARB process, a 50% rolled sheet is cut into two, stacked to the initial thickness, and then rolled again. As is evident, the rolling in the ARB is not only a deformation process but also a bonding process (roll bonding). To achieve good bonding, the surface of the materials is degreased and wire brushed before stacking, and the roll bonding is sometimes carried out at elevated temperatures below the recrystallization temperature of the materials. Repeating this ARB process a number of times can lead to very large plastic strain in the materials.

It has been observed that achieving good bonding is the critical step in the ARB process. Surface treatment is indispensable for achieving good bonding. Furthermore, there is a critical rolling reduction in one-pass roll bonding, below which it is difficult to achieve sufficient bonding. Though the critical reduction depends on the materials and the processing temperatures, more than 35% reduction by one pass is necessary in general, so that the

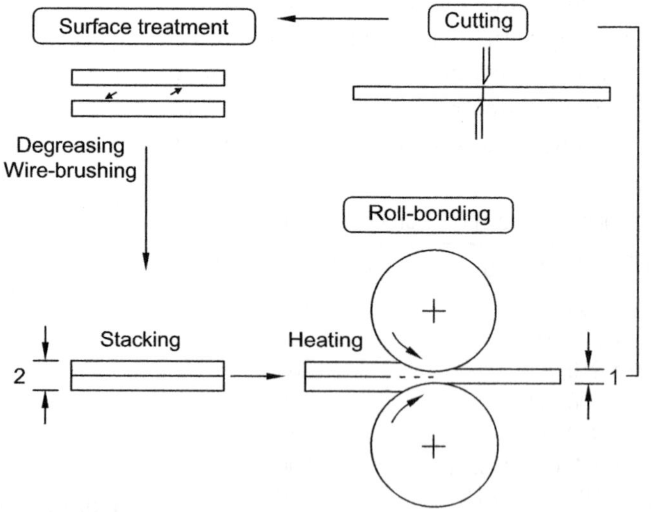

Fig. 3.22 Principle of accumulative roll bonding.

rolling force is large in comparison with conventional rolling. Except for the necessity of large enough capacity of the rolling mill which can realise one-pass heavy roll bonding, there are no special requirement in the equipments for the ARB. A serious problem in the ARB process is fracture of the materials. Since a large amount of total plastic strain is accumulated in the materials, the rolling is not a hydrostatic process; edge-cracks sometimes appear in the sheets, especially at higher cycles. In certain kinds of materials, such as Al–Mg alloy, the edge cracks largely propagate into the centre of the sheets. In that case, it becomes impossible to proceed to the subsequent cycles. With ductile materials, for example, pure aluminum and iron, the UFG sheets with dimension 1 × 50 × 300 mm are fabricated without cracking by the ARB process.

Irrespective of the kind of material studied, the average grain diameter of UFG materials is around 500 nm. The UFGs are surrounded by clear but irregular-shaped boundaries; the number of dislocations inside the grains seems small. These features are similar to those observed in materials that have been heavily deformed by other SPD processes. The most characteristic feature of the UFGs in the ARB-processed materials is the elongated morphology. Measurements clearly confirm that the elongated UFGs are not subgrains, but are grains surrounded by high-angle grain boundaries.

ARB-processed materials with elongated UFG structures exhibit very high strength. The grain size and the tensile strength of the various UFG materials fabricated by ARB are summarised in Table 3.5. In most cases, the mean grain thickness of the pancake-shaped UFGs or the ultrafine lamellar structures are 100 ~ 200 nm. Materials with higher purity tend to show larger grain size. ARB at lower temperature results in smaller grain size within the similar materials. UFG materials exhibit tensile strength which is two to four times higher than that of the starting materials with conventional grain sizes. On the other hand, ARB-processed materials have limited tensile elongation owing to early plastic instability.

Table 3.5 Grain size and tensile strength of various ARB-processed metals and alloys

Materials (wt.%)	ARB process	Grain size (nm)	Tensile strength (MPa)
4N-Al	7 cycles RT	670	125
100-Al (99%Al)	8 cycles RT	210	310
5052-Al (Al–2.4Mg)	4 cycles RT	260	388
5083-Al (Al–4.5Mg-0.6Mn)	7 cycles at 100°C	80	530
6061-Al (Al–1.1Mg–0.6Si)	8 cycles RT	100	357
7075– I (Al–5.6Zn–2.6Mg–1.7Cu)	5 cycles at 250°C	300	376
OFHC-Cu	6 cycles RT	260	520
Cu–0.27Co–0.09P	8 cycles at 200°C	150	470
Ni	5 cycles RT	140	885
IF steel	7 cycles at 500°C	210	870
0.041P-IF	5 cycles at 400°C	180	820
SS400 steel (Fe–0.13C–0.37Mn)	5 cycles RT	110	1030
Fe–36 Ni	7 cycles at 500°C	87	780

3.2.5 Nanolithography

STM-based nanolithography This has been exploited for local oxidation and passivation, localized chemical vapour deposition, electrodeposition, mechanical contact of the tip with the surface, and deformation of the surface by electrical pulses. Patterns with a minimal size from 10–20 nm to 1 nm in ultrahigh vacuum (UHV) have been demonstrated. Nanometre-sized holes can be formed using low energy electrons from a scanning tunnelling microscope (STM) tip when a pulsed electric voltage is applied in the presence of sufficient gas molecules between the substrate and the tip. For example, holes that are 7 nm deep and 6 nm wide on highly ordered pyrolytic graphite (HOPG) substrate were formed in nitrogen at a pressure of 25 bar by applying a 7 V pulse to the tip for 130 mins, with the distance between the tip and the substrate being 0.6–1 nm. A possible mechanism is that the electric field induces the ionisation of gas molecules near the STM tip, and accelerates the ions towards the substrate. Ions bombard the substrate and, consequently, nanometre-sized holes are created. A certain electric field is required to generate field emitted electrons. The diameter of an electron beam ejected from an STM tip is dependent on the applied bias voltage and the diameter of the tip. At low bias (<12 V), the diameter of the ejected electron beam remains almost constant; however, the beam diameter changes significantly with bias voltage and the diameter of the tip.

Nanostructures can be created using field evaporation by applying bias pulses to the STM tip–sample tunnelling junction. For example, nanodots, nanolines, and nanocorrals of gold on a clean stepped Si (111) surface were fabricated by applying a series of bias pulses (< 10 V and ~30 μs) to an STM gold tip at UHV (a base pressure of 10^{-10} mbar). Nanodots with diameter as small as a few nanometres can be realised. By decreasing the distance between adjacent nanodots, it was possible to create continuous nanolines, a few nanometres wide and over a few hundred nanometres long. A nanocoral of diameter about 40 nm, formed by many Au nanodots, each with diameter of a few nanometres, was also created on the Si (111) surface.

AFM-based nanolithography Direct contacting, writing or scratching is referred to as a mechanical action of the AFM tip that is used as a sharply pointed tool in order to produce fine grooves on sample surfaces. Although direct scratching creates grooves with high precision, low quality results are often obtained due to tip wear during the process. An alternative approach is to combine scratching on a soft resist polymer layer, such as PMMA or polycarbonate, as a mask for the etching process and subsequent etching to transfer the pattern to the sample surface. This method ensures reduced tip damage, but also precludes an accurate alignment to the structures underneath. A two-layer mask has been investigated as a further improvement. For example, a mask coating consisting of a thin layer of polycarbonate of 50–100 m and a film of an easy-to-deform and fusible metal such as indium or tin was used to create 50 nm–wide structures. Figure 3.23 is a typical layout of the sample and process steps with AFM lithography.

Dip-pen nanolithography In dip-pen nanolithography (DPN), the tip of an AFM operated in air is 'inked' with a chemical of interest and brought into contact with a surface. The ink molecules flow from the tip onto the surface similar to a fountain pen. The water meniscus that naturally forms between the tip and the surface enables the diffusion and transport of the molecules, as shown in Fig. 3.24. Inking can be done by dipping the tip in a solution containing a small concentration of the molecules, followed by a drying step (e.g., blowing dry with compressed difluoroethane). Line widths down to 12 nm with spatial resolution of 5 nm have been demonstrated with this technique. Species patterned with DPN include conducting polymers, gold, dendrimers, DNA, organic dyes, antibodies and alkanethiols. Alkanethiols have also been used as an organic monolayer mask to etch a gold layer, and subsequently etch the exposed silicon substrate.

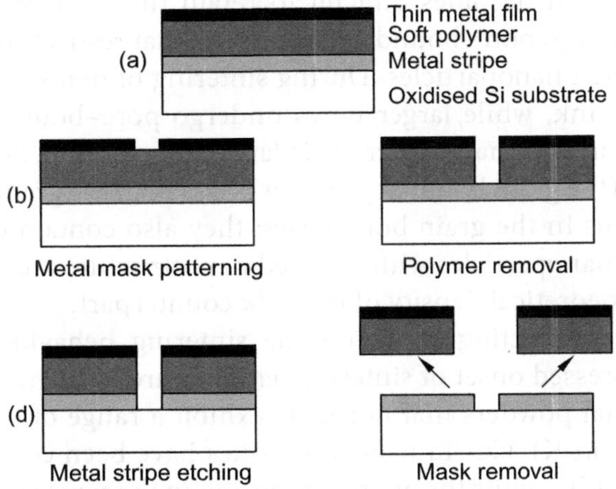

Fig. 3.23 Layout of the sample and the process steps with AFM lithography: (a) sample multilayer structure, (b) thin mask patterning by AFM lithography, (c) polymer removal in plasma oxygen, (d) titanium stripe etching, and (e) resulting electrodes after sacrificial layers removal.

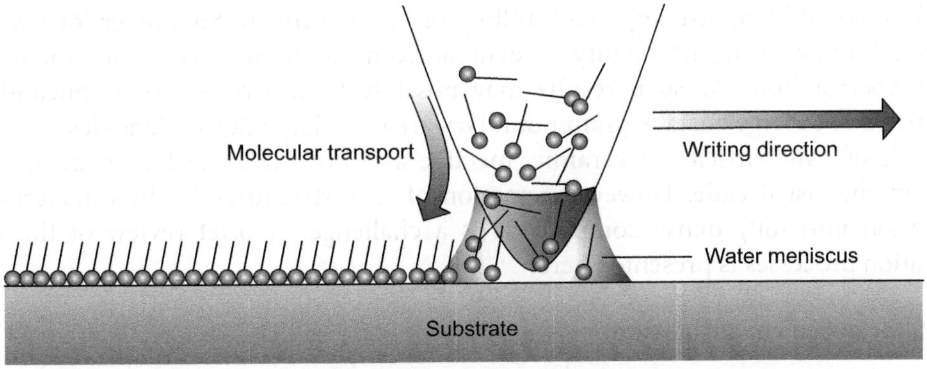

Fig. 3.24 Schematic representation of the working principle of dip-pen nanolithography.

3.3 CONSOLIDATION OF NANOPOWDERS

The commercial application of nanomaterials beyond the boundaries of materials science laboratories is possible only on successful consolidation of these materials into bulk-sized components preserving the nanostructures. Due to the long duration of sintering at high temperature that is involved in conventional consolidation techniques, it is difficult to retain the nanograin-size due to grain growth in such techniques.

The density of the green compact depends on the frictional forces of the powder particles that originate from electrostatic, van der Waals and surface adsorption forces. These forces are significantly high in nanoparticles, forming hard agglomerates and inter-agglomerates, which are relatively large. Further, nanoparticles contain a large number of pores which require not only higher temperature but also prolonged sintering times for their successful elimination; consequently, it becomes difficult to retain the grain size in the nanometre domain. Large pores undergo pore–boundary separation that restricts the attainment of full density in the consolidated nanoparticles. During sintering of nanoparticles, pores smaller than the critical size shrink, while larger pores undergo pore–boundary separation. The fraction of grain boundaries in nanomaterials is large compared to that in coarse-grained materials. The density of the grain boundary regions is less than the grain interior due mainly to the relaxation of atoms in the grain boundaries; they also contain other lattice defects. Therefore, consolidated nanoparticles with retained nanostructure are expected to exhibit a density lower than the theoretical density of the bulk counterpart.

There are numerous conflicting views on the sintering behaviour of nanoparticles. Nanoparticles show depressed onset of sintering temperature to the range of 0.2 T_m–0.3 T_m compared to conventional powders that normally exhibit a range of 0.5 T_m–0.8 T_m, where T_m is the melting point (in K). Cu–Ta nanocomposites have been consolidated to close to theoretical densities at 500°C, while the microcomposites in the same system need a sintering temperature of 1000°C (Fig. 3.15). Similarly, W, which is used in filaments, is usually sintered at about 2800°C due to its high melting point. However, nanocrystalline W could be sintered to close to its theoretical density at 1500°C itself leading to significant energy savings. Figure. 3.25 shows the dilatometry results of microcrystalline W and nanocrystalline W (50 mm) obtained by high-energy ball milling of W (50 mm) in Simoloyer for two hours. Significant differences in sinterability are evident from the extent of consolidation observed in the nanocrystalline W. Such results may possibly be attributed to significantly high diffusivities due to large surface/grain boundary area and large defect densities.

Synthesis of nanoparticles of ceramic, metallic and their mixture has made substantial progress in the last decade. However, retention of nanostructures in these materials after consolidation into fully dense components is a challenge. A brief review of the various consolidation processes is presented here.

3.3.1 Shockwave consolidation

Shock wave consolidation, also termed *dynamic consolidation*, is used to densify powdered materials without inducing thermal activated microstructural and compositional changes

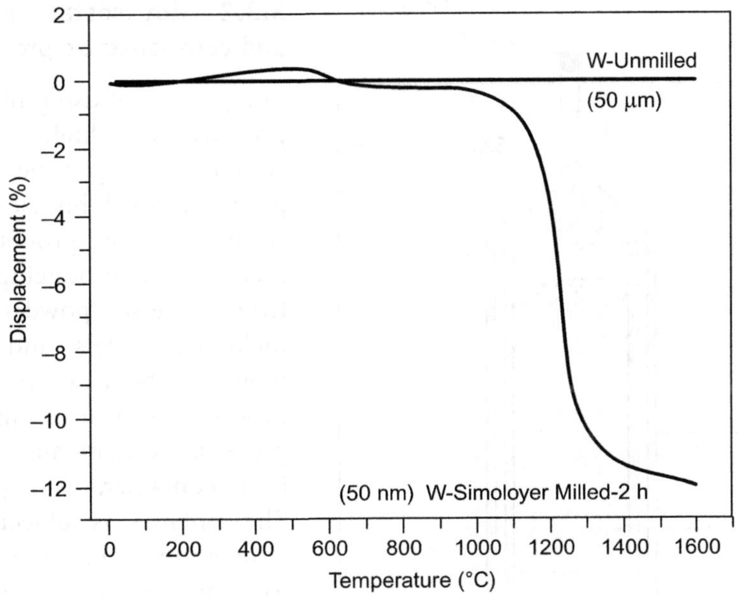

Fig. 3.25 Improved sinterability of nanocrystalline W in comparison to microcrystalline W. (*Source:* BS Murty, IIT Madras).

that are normal in conventional thermomechanical processing. Such densification is possible because of the interparticle bonding due to localized melting at the interfaces between the particles. High pressure and rapid loading rates that create high plastic deformation finally lead to high shock initiated chemical reactions completely different from the conventional ones. Metallic or alloy powders are usually processed through this route due to the fact that plastic deformation in metal is comparatively easier than for ceramics.

In this process, particles are enclosed in a steel block that looks similar to a cold compaction chamber that is covered using a plate, on top of which a driver plate usually bears the brunt of explosions and drives the shock wave towards the sample (Fig. 3.26). The driver plate is usually made of highly conductive and ductile material. Explosives are packed on top of this plate carefully and the spillage, if any, has to be removed. The detonator is set on top of the explosive and the set up is ready for processing to start. Ammonium nitrate is usually used as the explosive. Shock compaction of powders is a dynamic consolidation technique which provides a viable method for densification of amorphous as well as ultrafine powders. Fe-, Ni- and Ti-based alloys have been studied more often using this technique. Pure metals and alloys with retained nanostructures can be consolidated using this technique. Some of the systems that have been investigated are: $Fe_{73.5}Cu_1Nb_3Si_{13.5}B_9$, Fe, Fe–Al, diamond–Si, Al, $Pr_2Fe_{14}B/\alpha$-Fe, Ti_5Si_3 and NiAl.

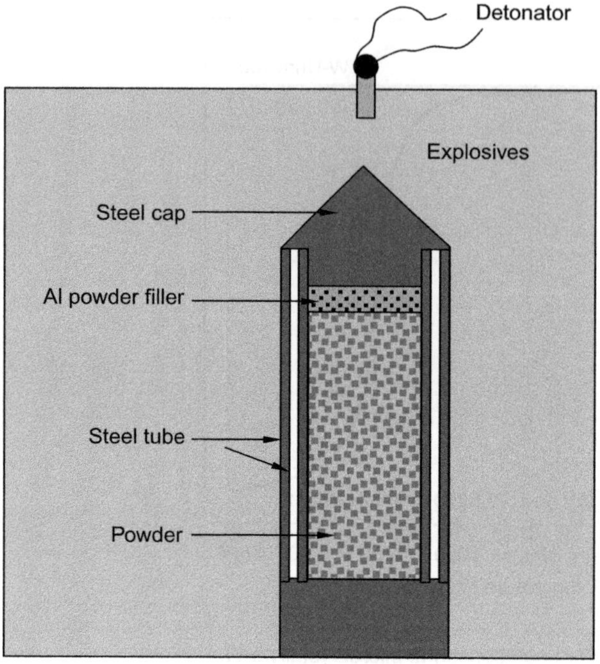

Fig. 3.26 Schematic of shock wave consolidation setup.

3.3.2 Hot isostatic pressing (HIP) and cold isostatic pressing (CIP)

The process of using high hydrostatic pressure and high temperature to compress fine particles into coherent parts is termed 'hot isostatic pressing' (HIPing). This process provides a method for producing components from diverse powdered materials, including metals and ceramics. A typical schematic of the HIPing process is shown in Fig. 3.27. In this process, the elemental blend is placed in a container, usually a steel can. The container is subjected to elevated temperatures, and very high vacuum to remove air and moisture from the powder, after which it is sealed and HIPed. The application of high inert gas pressures and elevated temperatures results in the removal of internal voids and creates a strong bond throughout the material. The result is a clean homogeneous material with a uniformly fine grain size and near 100% density.

In cold isostatic pressing (CIPing), pressure is applied from multiple directions to get better uniformity of compaction, compared to uniaxial pressing. There are two methods in isostatic pressing. In wet bag isostatic pressing, the powder is encased in a rubber sheath that is immersed in a liquid that transmits the pressure uniformly to the powder. In dry bag isostatic pressing, instead of immersing the tooling in a fluid, the tooling itself is built with internal channels into which high pressure fluid is pumped.

Some of the material systems that have experimented with the HIPing process are TiC, TiCN, SiC, Si_3N_4–Y_2O_3–Al_2O_3, Si_3N_4, intermetallic TiAl, TiAlSi, $MgAl_2O_4$, Ni–Fe, Y_2O_3 dispersed steels, $MgAl_2O_4$ dispersed steels, Al and carbon nanotube reinforced Si_3N_4. There have been instances where HIP is used as an add-on processing to CIP to increase the densification and thereby mechanical properties, for example, consolidation of Fe. Using CIP as a primary consolidation technique works well for a micron-sized particle whereas a nanoparticle might need some kind of secondary consolidation process to form a bulk part, as the nanoparticles are friable and can crumble even if crushed with a hand. The only solution is to either use sintering after CIPing or to use HIPing. It has been found that HIPing leads to better densification in less time compared to the time required for sintering to achieve the same density.

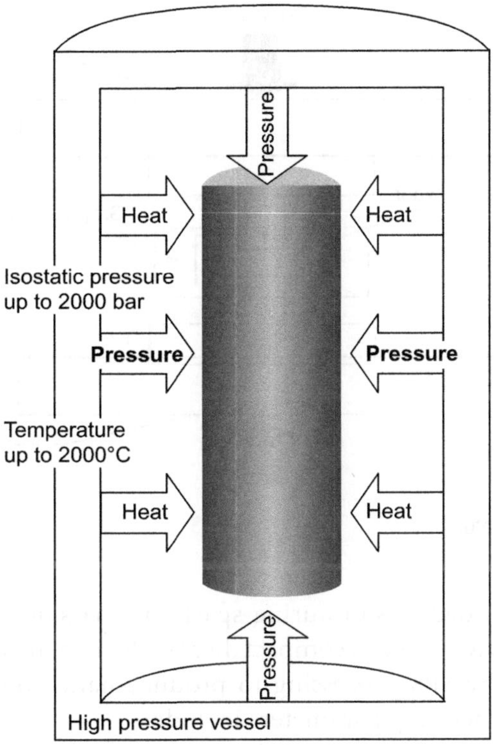

Fig. 3.27 Schematic of a hot isostatic pressing setup.

3.3.3 Spark plasma sintering

With the growing need for processing and consolidating nanopowders, retaining the initial microstructure in the fabricated component remains a challenge. Though the manufacture of novel materials is limited to laboratory scale, vigorous demand in the market persists and is driving the demand for new consolidation techniques apart from hot pressing, high temperature extrusion and hot isostatic pressing. Since longer processing times at high temperatures often result in extraneous grain growth, the new nanopowder processing technique, namely spark plasma sintering (SPS), which has evolved in the last decade, overcomes these obstacles by applying rapid heating to sintering temperatures, leading to rapid sintering of nanocomposites. The attractive properties of nanopowders are well conserved by the fabrication of bulk material through SPS. Consolidation of metals, composites, ceramics, intermetallics, cermets, nanocomposites and carbon nanotube reinforced ceramics has been accomplished by spark plasma sintering.

The schematic of a typical SPS apparatus is shown in Fig. 3.28. It consists of a graphite die where powder is loaded and heated by passing an electric current. Hold time, ramp rate, pulse duration, and pulse current and voltage can be set for controlling the temperature of the sintering cycle. Change in temperature, pulse power (voltage/current), sintering displacement

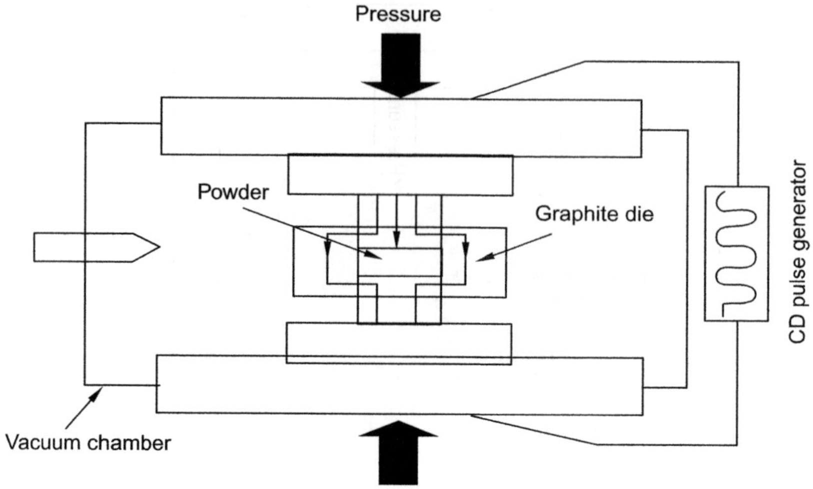

Fig. 3.28 Schematic of SPS setup.

and sintering rate are recorded in situ during spark plasma sintering. Powder mixtures are initially cold pressed (~ 200 MPa) into a compact disc (with diameter around 2 cm and thickness around 5 mm), followed by SPS processing to produce nano-consolidated nanocrystalline composites. Typical SPS processing parameters include:

- Applied pressure between 50 and 100MPa
- Pulse duration of ~ 10 ms with on–off cycle of 2–2.5 ms
- Maximum pulse parameter of 10,000 A and 10 V

After applying the given pressure, samples are heated to the preset temperature (for a few minutes) and are ramped rapidly (~150–500 K/min) to sintering temperatures with hold time of 3–5 min to complete the sintering. Samples are usually cooled to below 100°C within 5 minutes of the completion of sintering.

Densification of the green compact occurs in four stages: vacuum creation, pressure application, resistance heating and cooling down as shown schematically in Fig. 3.29. SPS is carried out in vacuum to account for removal of gases, and results in a densely consolidated composite. Heating is accomplished by spark discharge between particles, which activates the surface by removing surface oxide. This helps in heat and mass transfer between the purified particles to carry heat and sinter them to full density. Initially, samples show an increase in volume due to thermal expansion, followed by shrinkage occurring primarily during resistance heating when the green discs are held at high temperatures and pressure. Onset of sintering is dependent on the initial particle size, which decides the onset of shrinkage. It is generally difficult to determine when the sintering process is complete; however, shrinkage of the specimen could help in deducing the completion of SPS sintering. Some of the nanopowder material systems that have been consolidated with SPS process are amorphous Si–C–N, Ni–Ti shape memory alloy, $(Al–12.5Cu)_3Zr$ intermetallic, TiCN cermet, $SiC–ZrO_2–Y_2O_3–Al_2O_3$

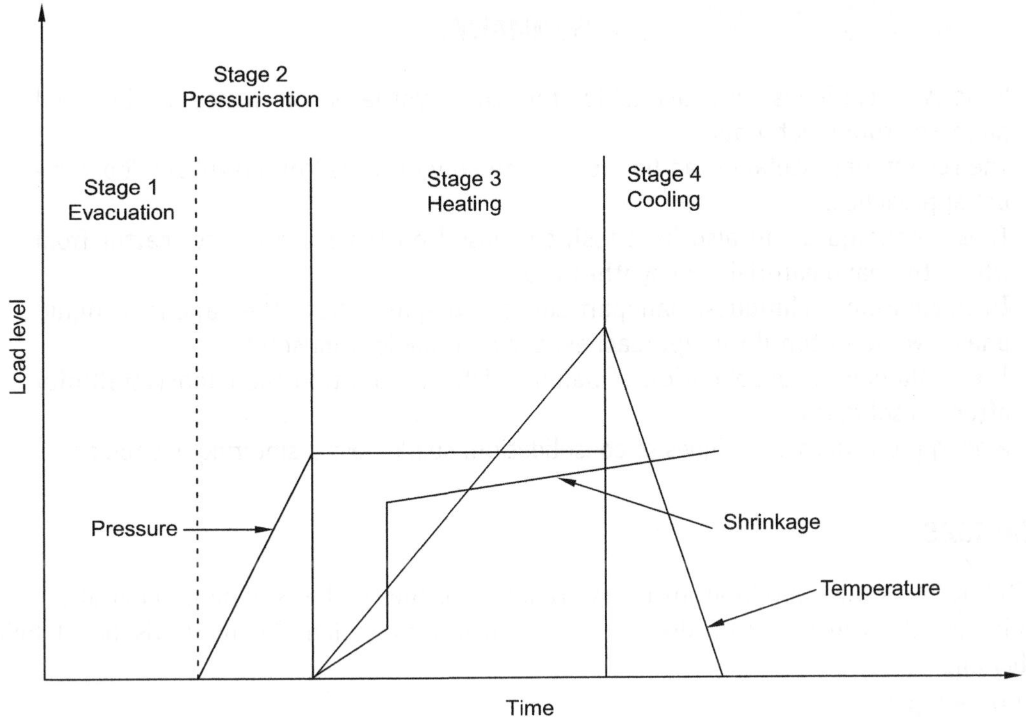

Fig. 3.29 Schematic of SPS sintering stages.

composites and CNT-Al_2O_3 composites. Figure 3.30 shows the TEM images of SPS consolidated structural nanocomposites (NiAl–Al_2O_3 and FeAl–Al_2O_3), both of which demonstrate that the nanocrystalline nature is retained after SPS.

Considering the rapid advancement in the field, it is a mammoth task to cover all the techniques of synthesis and consolidation of nanoparticles and nanocrystalline materials that are available, and the reader is encouraged to further explore this exciting world.

Fig. 3.30 TEM images of SPS consolidated samples of (a) NiAl–Al_2O_3 and (b) FeAl–Al_2O_3 nanocomposites; (c) Higher magnification image of a region in (b). (*Source:* BS Murty, IIT Madras).

SUMMARY

- Various techniques are available for the synthesis of nanoparticles and nanostructured materials.
- The techniques available can be grouped broadly into the 'top-down' and 'bottom-up' approaches.
- These techniques can also be classified, based on the state of the matter from which the nanomaterials are synthesized.
- In bottom-up techniques, nanoparticles are prepared from the vapour or liquid phase, while in top-down approaches, they are made from solids.
- The challenge in consolidation of nanoparticles is to retain the nanocrystallinity after consolidation.
- Among the various techniques of consolidation, spark plasma sintering is popular.

EXERCISES

1. Define 'top-down' and 'bottom-up' approaches for the synthesis of nanomaterials.
2. Identify the category (top-down or bottom-up) to which the methods listed below belong:
   (a) Sol–gel
   (b) MBE
   (c) E-Beam lithography
   (d) ECAP
   (e) Wire explosion
   (f) Laser ablation
3. Describe the principle involved with suitable figures in the following techniques:
   (a) RF and DC sputtering
   (b) Electron beam evaporation
   (c) Microwave plasma based CVD
4. It is desired to synthesize yttria nanoparticles of about 2–3 nm in size. Discuss the suitability of the following techniques for the purpose with reasoning.
   (a) Mechanical milling, (b) Electron beam evaporation, (c) CVD, (d) Sol–gel
5. Identify a suitable technique for the synthesis of (a) Au nanoshells and (b) Silica nanoshells, and list a few applications for each.
6. Describe the distinctive features of a self-assembled nanostructure.
7. Compare ion and electron beam nanolithography for nanofabrication.
8. List a few deformation-based techniques for the development of nanostructures.
9. Discuss briefly what we can learn from nature for the synthesis of nanostructures.
10. Identify techniques for the development of bulk nanostructured solids from nanopowders.

Chapter 4

Applications of Nanomaterials

Learning objectives

- Nano-electronic devices, MEMS, NEMS and sensors
- Use of nanoscience in the medical, food and agriculture industries, automobile, textile, water treatment and civil applications
- Application of nanotechnology for strategic use in energy, space and defense

Historically, there are several recorded instances of technologies that have revolutionized human civilization. From the invention of automobile wheels to the printing press, technological revolutions have resulted in remarkable improvement in the quality of life and have eventually led to societal transformations. With nanotechnology promising to impact almost every sector (Fig. 4.1), it is popularly believed that this could be the next revolution.

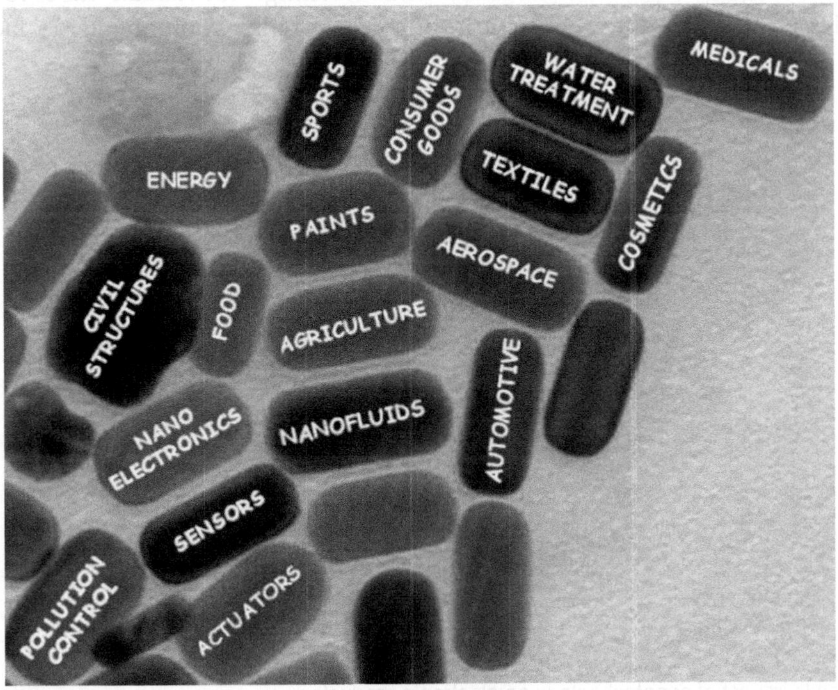

Fig. 4.1 Application potential of nanotechnology in diverse fields.

Since nanomaterials possess unique chemical, physical and mechanical properties, they can be used for a wide variety of applications—from toothpaste to satellites. Nanotechnology is being used in virtually all fields, from science to engineering, and health to medicine. Although nanotechnology is still only in its infancy, the consumer world is already exploding with 'nanotechnology enhanced' products. The revolution in communications systems, represented by the common sight of everyone, including the young and old, student and professional, craftsman and scientist, holding a mobile telephone, is perhaps the most obvious evidence of the potential of this new technological imperative. Nanomaterials are also used in cosmetics, textiles, healthcare, tissue engineering, catalysis, functional coatings, medical diagnosis and therapeutics, sensors and communication engineering, and water and air pollution treatment. Chapter 1 gave a brief description of some of the applications of nanotechnology. This chapter will provide a brief description of the applications of nanotechnology in a wide spectrum of fields.

4.1 NANO-ELECTRONICS

Arguably, the most intangible impact of nanotechnology has been in the field of electronics (Fig. 4.2). The last few decades have witnessed a dramatic decrease in feature size accompanied by an enhancement in processing speed. Semiconductor electronics technology has seen a revolutionary change—the transition from macroscopic to nanoscale transistors.

In 1897, JJ Thompson discovered electrons while working with vacuum tubes to investigate the properties of cathode rays. The vacuum tubes were bulky components. In fact, the first

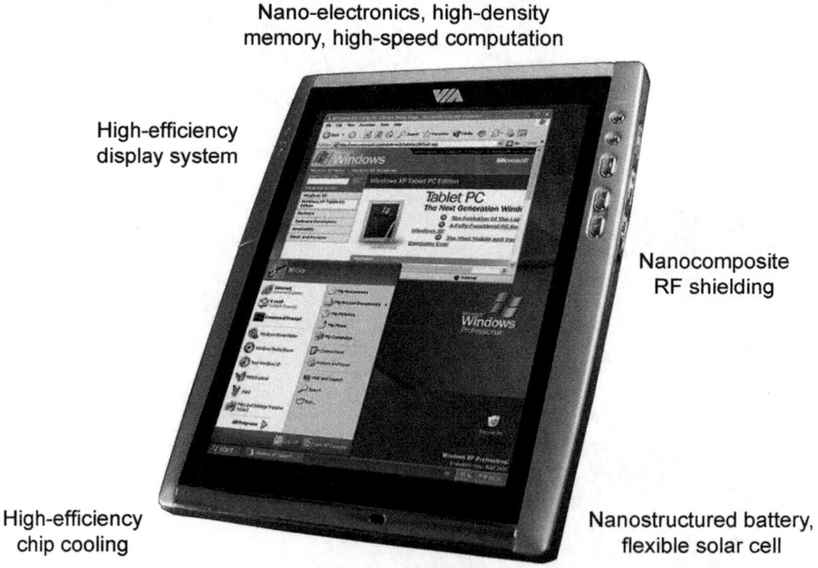

Fig. 4.2 Possible impact of nanotechnology on electronic products.
(*Source:* http://commons.wikimedia.org/wiki/File:VIA_Tablet_PC_Reference_Design.jpg)

computers to be built in the 1940s had over 10,000 vacuum tubes and occupied over 93 square metres of space. The main limitation of vacuum tubes, apart from the huge space requirements, was the associated high heat dissipation, poor efficiency, slow speed and poor reliability of vacuum tube amplifiers.

Soon researchers turned to solid-state electronic materials to circumvent the limitations of vacuum devices. The development of semiconductors and advances in their fabrication techniques, including zone refining, doping by thermal diffusion and lithography, have been the key to the large-scale development of novel electronic devices. In most applications, thermionic devices (vacuum tubes) have been replaced by semiconductor devices. In semiconductor devices, electronic conduction in the solid state is used as compared to conduction in the gaseous state or thermionic emission in vacuum tubes. The invention of transistors in 1947 by John Bardeen and Walter Brattain at Bell Laboratories heralded a greater revolution in electronics, virtually redefining the lifestyle of many. This transistor was initially based on a point contact configuration. Soon after, Shockley demonstrated the first field effect transistor (FET) based on n–p–n junctions that are used even today with suitable modifications.

The distinct advantage of transistors was that they could work as amplifiers and as switches. They were also much smaller than vacuum tubes and consumed much less power. This enabled the design of more complex digital circuits with higher switching speeds and efficiency. However, it was soon realised that circuits based on individual transistors were large and difficult to assemble. There was also significant time lag in response due to the long distances the current had to travel. In order to increase the speed by reducing the delay time, there was a need for miniaturization of electronic components and a technology for integrating a large number of them with high precision to enable realization of sophisticated digital circuits.

This eventually led to the development of integrated circuits (IC). The first IC board was manufactured in 1958 and 1959 by Jack Kilby at Texas Instruments, and Robert Noyce at

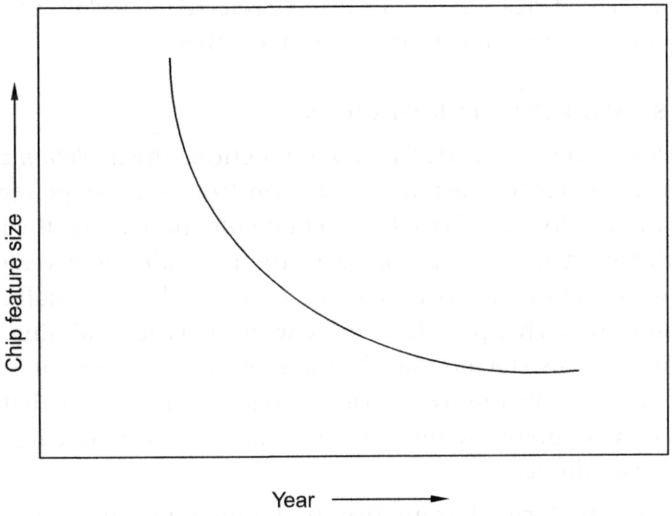

Fig. 4.3 Schematic representation of Moore's law on chip feature size.

Fairchild Camera. With this, people were able to make several transistors on the same semiconductor. Not only transistors, but other electronic components such as resistors, capacitors and diodes could also be made by the same process and with the same materials.

In 1965, Moore predicted that the number density of components that could be fabricated on an integrated circuit at an affordable cost would roughly double every 18 months. The simplest formulation of this law is that the number of transistors in an IC doubles every 18 months. This profound statement, now referred to as Moore's law (Fig. 4.3), has been found to hold true at least till date, i.e., for about four decades since its statement. The law itself was not based on any derivation, but rather on statistical analysis and scientific intuition. This miniaturization has also been associated with an increase in the efficiency, memory and processing speed of digital devices. The cost per transistor has reduced, although the cost per unit area has increased due to increase in processing costs.

4.1.1 Fundamentals of semiconductor devices

The electronic properties of semiconductor materials can be tailored significantly by careful control of processing parameters and chemistry, by manipulating the addition of minor impurities called dopants. This is perhaps the main motivator for the widespread application of semiconductor materials in the electronics industry. The conductivity of a semiconductor can be changed by electric field, light, pressure or heat, and hence they can be excellent sensors. The charge carriers responsible for current conduction in a semiconductor can be either electrons (negatively charged) or holes (positively charged). Doping a semiconductor (for example, silicon) with small amounts of impurity atoms (for example, phosphorus or boron) greatly increases the concentration of charge carriers and thereby changes its conductivity. Doped semiconductors with excess holes are called 'p-type semiconductors', and those with excess free electrons are called 'n-type semiconductors'. These semiconductors are doped under controlled conditions in a fabrication facility, usually referred to as *fab*. The p–n junctions are those where n-type and p-type semiconductors join together.

Conventional semiconductor devices

Diode The *p–n junction diode* is made from a p–n junction. The *depletion zone* formed at the p–n junction acts as a barrier for current conduction from the n-type region to the p-type region. When the device is forward biased, current conduction across the junction occurs as the p-side is at a higher electric potential compared to the n-side. However, on reverse biasing, the depletion zone which acts as a barrier to conduction results in a small current value only. Thus the characteristics of such a p–n diode vary with the type of biasing. The behaviour of semiconductor devices can also be modified by the external environment, such as exposure to light, which increases the number of free carriers and hence the conductivity. Such diodes are known as *photodiodes*. Compound semiconductor diodes can generate light, similar to light-emitting diodes and laser diodes.

Transistor Transistors are formed from two p–n junctions, in either n–p–n or p–n–p configuration. The region between the junctions (*base*) is usually extremely narrow, with the *emitter* and the *collector* acting as the terminals. The device characteristics can be changed by injecting a small current through the junction between the base and the emitter, such that the transistor can conduct current even though it is reverse biased. This increases the magnitude of current flowing between the collector and emitter, and can be nicely controlled by the base–emitter current. The field effect transistor works on the principle that an electric field can increase or decrease the conductivity of the semiconductor. Application of an electric field leads to an increase in conductivity due to increase in the concentration of charge carriers. There are typically two common methods to apply the field, namely, junction field effect transistor (JFET) and metal oxide semiconductor field effect transistor (MOSFET). The field is applied by a reverse-biased p–n junction in a JFET. Conversely, in MOSFET, the field is generated using an electrode (gate) that is electrically isolated from the bulk material by using an intermediate insulating oxide layer. Among the semiconductor devices, MOSFET is the most widely used—*n-channel* (for electrons) and *p-channel* (for holes) are the two types of MOSFETs.

SEMICONDUCTOR DEVICE MATERIALS

Silicon (Si) is the most popular semiconducting material for device fabrication. This is because, among all semiconducting materials, it is available in large quantities, its raw materials are inexpensive and it is easy to process. Silicon is currently fabricated as wafers to sizes as large as 300 mm. Though germanium (Ge) is also a widely used semiconductor material, it is not as popular as Si due to its thermal sensitivity. Germanium is usually alloyed with silicon to make high-speed SiGe devices. Although gallium arsenide (GaAs) can also be used in high-speed devices, it cannot be manufactured into large sizes. Thus, mass production of GaAs devices is significantly more expensive than that of silicon.

Silicon carbide (SiC) is used for blue light emitting diodes (LEDs) and is used in semiconductor devices, which operate at high temperatures and in environments containing ionising radiation. Indium compounds such as indium arsenide, indium antimonide and indium phosphide are also used in LEDs. Selenium sulphide is being used for photovoltaic solar cells.

Currently there are several components fabricated on a nanometre scale. Apart from miniaturization of microelectronic components, development of nano-interconnects is also a challenging area. Here, carbon nanotubes and self-assembled metallic or organic structures are being investigated. Besides miniaturization of conventional FETs, the ability to synthesize electronic materials on the nanoscale with required precision has also resulted in the development of novel electronic devices with quantum mechanical behaviour. These include molecular diodes, single electron devices and tunnelling devices.

4.1.2 Metal oxide semiconductor field effect transistor (MOSFET)

Transistors can perform both as switches and as amplifiers. As switches, they can be used as a two-state device, representing a binary 1 or 0 state in digital circuits. Currently, MOSFETs are among the most popular type of transistors used. A field effect transistor typically consists of three terminals referred to as source, drain and gate. In MOSFET, a metal electrode separated from a semiconductor below by an insulating thin oxide barrier functions as the gate (Fig. 4.4). At low voltages, not enough carrier charges are able to build below the gate, i.e., between the source and the drain. However, as the gate voltage is increased, more electrons are attracted below the gate and current flow between the source and drain increases dramatically.

Although individual MOSFETs with gate length as low as 40 nm have been demonstrated, their reliability and reproducibility for industrial-scale production is still uncertain. It is expected that the minimum achievable gate length could be about 70 nm, setting the limit perhaps for component density in ICs. A few constraints for further scaling down are listed below:

- Application of high (gate) bias voltage over short distances can lead to avalanche breakdown;
- As the number of components per unit area increases with miniaturization, the adverse effects of heat dissipation of transistors on closely spaced components become a major factor in the performance of the devices;
- Effect of microscopic heterogeneities in dopant distribution can become more predominant as device dimension decreases.

To overcome these limitations, solid-state quantum effect devices were synthesized.

4.1.3 Solid-state quantum effect devices

The physical dimension of quantum devices are such that there is confinement of electrons in at least one of the axes. They can thus be classified into one of the three following categories:

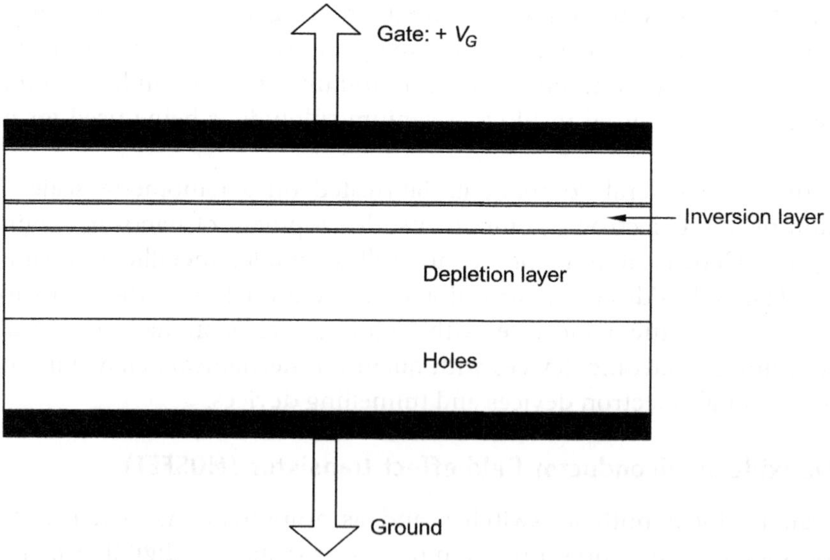

Fig. 4.4 Schematic of MOSFET.

Quantum dots (QDs) These are dot-like islands of semiconductors with electron confinement along all three directions. This results in zero classical degrees of freedom as electronic states are quantised in all three dimensions.

Resonant tunnelling devices These are 2D quantum devices and consist of a long and narrow semiconductor island, with electron confinement only in two directions. The wire/tube is too long in at least one direction to display quantum properties along that axis. These devices have one or two classical degrees of freedom. Depending on the applied bias between the source and drain, a tunnelling current can be effected in a resonant tunnelling diode (RTD).

In contrast to RTDs, the resonant tunnelling transistors (RTT) work on a three-terminal configuration, with the gate voltage deciding the current flowing through the device.In contrast to conventional MOSFETs, which can perform only as two-state switches, RTTs and RTDs can display multiple on-and-off states associated with multiple discrete quantum levels inside the potential well on a very small or very narrow island. Hence, it would be possible to achieve the same logic using fewer devices in the circuit. This can help decrease the problems caused by heat dissipation associated with increasing component density. Hence, solid-state quantum effect nano-electronic devices are likely to provide solutions to at least a few constraints posed on miniaturization of microelectronic components. It is expected that the limit for Moore's law, which is based on the minimum feature size of conventional MOSFET (70 nm), can be scaled down using solid-state quantum devices (25 nm).

Single electron transistors (SET) These devices are based on a metallic island structure with three degrees of freedom. These can be contrasted to QDs, with zero degrees of freedom. Although the physical size of the QD and SET can be similar (100 nm), QDs are quantised since they are synthesized of semiconductors, while 100 nm is too large to observe quantisation in metallic islands in SETs. SET devices possess a three-terminal configuration. Very small variations in the charge of the gate, amounting to a difference of one single electron charge, can result in the on-and-off switching function of SET; hence they are termed single electron transistors.

Compared to conventional microelectronic devices, composed of Si, quantum devices are fabricated from III–V semiconducting materials (like GaAs and AlAs), by virtue of their higher electron mobility and lower defect density at junctions.

4.1.4 Hybrid micro–nano-electronic resonant tunnelling transistors

Nanoscale RTTs tend to be difficult to fabricate with sufficient uniformity in large quantities because of their relatively complex structure, small size and the sensitivity of the effects they employ. However, attempts have been made to fabricate 'hybrids' of solid-state quantum effect devices combined with micron-scale transistors to optimise the benefits of such nanoscale quantum devices. One such hybrid transistor-like device, the RTD-FET, is constructed by building tiny, nanoscale quantum effect RTDs into the drain (or source) of a bulk-effect micron-scale FET. Such a hybrid RTT can exhibit multistate switching behaviour of the type described in the preceding section for purely nanoscale RTTs. For this reason, the hybrid device can represent more logic states than a pure bulk-effect, microelectronic FET on the integrated circuit.

Thus, the density of the logic can be increased using the multistate switching characteristics of hybrid micro–nano-electronic RTTs, without appreciably increasing the density of the devices on an integrated circuit. Also, these hybrid RTTs share the advantages of low power and high speed exhibited by the purely nano-electronic RTTs. Most importantly, fabricating circuits with this relatively large, hybrid-type of three-terminal RTT are easier than fabricating circuits with the much tinier, complex structures for purely nano-electronic RTTs.

4.1.5 Molecular electronic devices

Molecular electronic devices are usually synthesized from covalently bonded units that are electrically isolated from a bulk substrate. The major advantage of these systems is that they can be synthesized in billions through advanced chemical syntheses methods with a high degree of reproducibility and control, to ensure fabrication of devices with similar performance. Molecular devices can be classified into four major categories, namely:

1. *Electric field–controlled molecular switching devices*
2. *Electromechanical molecular electronic devices:* These involve the application of electrical or mechanical forces to move a switching molecule to turn the current on and off in the device.
3. *Photoactive/photochromic molecular switching devices:* These devices rely on the optical properties of a molecule to change their electron configuration, to enable the switching action of the device.
4. *Electrochemical molecular devices:* These are based on electrochemical reactions to enable the switching action of the device.

Individual molecular switching devices could be as small as 1.5 nm across, with densities of approximately 1012 devices/cm^2. It is postulated that this decrease in size could result in memory capacities of terabytes on a chip and in excess of one trillion switching devices on a single CPU chip. A primary advantage of molecular electronics is that molecules are natural nanoscale structures that can be made absolutely identical in vast quantities (approximately 10^{23} at one time).

Novel semiconductor devices

An example of such novel devices is based on *spintronics*. The dependence of the resistance of a material (due to the spin of the electrons) on an external field is called *magneto-resistance*. This effect can be significantly amplified (GMR: giant magneto-resistance) for nano-sized objects: for example, when two ferromagnetic layers are separated by a non-magnetic layer, which is several nanometres thick (for example, Co–Cu–Co). The GMR effect has led to an enormous increase in the data storage density of hard discs and has made the gigabyte range possible. The so-called tunnelling magneto-resistance (TMR) is very similar to GMR and is based on the spin-dependant tunnelling of electrons through adjacent ferromagnetic layers. Both the GMR and the TMR effect can be used to create a non-volatile main memory for computers, such as magnetic random access memory or MRAM.

Spintronics Spintronics ('spin-based electronics'), also known as magneto-electronics, is an emerging technology, which exploits the dual property of electrons, namely charge and spin state. The difference in the electron spin direction manifests as a weak variation in the magnetic energy state and is identified as the 'spin up' and 'spin down' states. Conventional solid-state semiconductor devices function in a purely binary proposition, where an electron's state or current represents only 0 or 1. Hence, a range of eight bits can represent every number between 0 and 255, but only one number at a time. In contrast, spintronics quantum bits (known as *qubits*) exploit the 'spin up' and 'spin down' states as superpositions of 0 or 1 with entanglement, so a register consisting of two spintronics qubits will have eight possible states instead of four.

Spintronic devices are used in mass storage devices. The storage capacity of hard drives has been following *Kryder's law*, which predicts doubling of the storage capacity almost every year. Thus, the storage capacity of hard discs appears to grow faster than the number of transistors in an integrated circuit as defined by Moore's law (doubling every 18 months). Present day hard disc drives use the spin effect, that is, the giant magneto-resistive effect. A spintronic device can generate a current of spin-polarised electrons, which can be changed based on the spin states.

Spin-polarised current can be easily generated, when current is sent through a ferromagnetic material. This effect is used in giant magneto-resistance devices. A GMR device usually has at least two layers of ferromagnetic materials that are separated by a spacer layer that is non-magnetic. An electrical current will flow easily, if the two magnetization vectors of the ferromagnetic layers are aligned, and if they are anti-parrallel, there will be resistance in the system. GMR has two variants, namely, current-in-plane, with the electric current flowing parallel to the layers and current-perpendicular-to-the-plane, with the current flowing perpendicular to the layers.

Spin valve is the most successful spintronic device thus far. In this device, a layered structure of thin films of magnetic materials is used, which leads to variation in electrical resistance with change in the direction of applied magnetic field. In case of spin valves, magnetization direction in one of the ferromagnetic layers is fixed, while that in the other layer is free to rotate. The electrical resistance of the device is lowest when the free layer and the pinned layer magnetization vectors are aligned. The electrical resistance of the device increases when the free layer magnetization vector rotates anti-parallel to the pinned layer magnetization vector due to magnetic field. The magnitude of the change is given by:

$$\frac{\text{Anti-parallel resistance} - \text{Parallel resistance}}{\text{Parallel resistance}} \times 100\%$$

This ratio is called the *GMR ratio*. GMR ratios as high as 200% have been realised in devices. In contrast, the anisotropic magneto-resistance effect in single-layer materials is usually less than 3%. Spin valves with magnetically soft free layers are very sensitive to weak fields, and have replaced anisotropic magneto-resistance sensors in computer hard disc drive heads since the late 1990s. MRAM devices also operate on spintronic principles.

4.1.6 Novel opto-electronic devices

In modern communication technology, traditional analogue electrical devices are being increasingly replaced by optical or opto-electronic devices due to their enormous bandwidth and capacity. Photonic crystals and quantum dots are good examples for this. Photonic crystals are defined as materials in which the refractive index varies periodically, whose lattice constant is half the wavelength of the light used. They are in one way similar to semiconductors and have selectable band gap light or photons instead of electrons.

Quantum dot lasers

Quantum dots are nanoscale objects which can be used, among many other things, for the construction of lasers. In a quantum dot laser, the emitted wavelength is a function of the diameter of the dot. These are inexpensive in comparison to conventional laser diodes and provide a higher beam quality.

Photonic crystals

Photonic crystals are periodic dielectric or metallo-dielectric (nano) structures that are designed to affect the propagation of electromagnetic waves (EM) in the same way as the periodic potential in a semiconductor crystal affects electron motion by defining allowed and forbidden electronic energy bands.

If the photonic crystal has to operate in visible light, its periodicity should be half of the wavelength of light. In order to make these in a simpler way, there have been attempts to grow photonic crystals as self-assembled structures from colloidal crystals. Opal, a gemstone, is a naturally occurring photonic crystal. The colours displayed by opal are due to Bragg diffraction of light from the lattice planes, which is a photonic crystal phenomenon. Such photonic crystals are also observed on the wings of some butterflies.

Photonic crystals can be used as optical materials for controlling the flow of light. Two-dimensionally periodic photonic crystals have already reached a level where integrated device applications are in sight, whereas their 3D counterparts are still far from commercialisation but will offer additional advantages, possibly leading to new device concepts, when some technological aspects such as manufacturability and principal difficulties such as disorder are under control. The first commercial products involving two-dimensionally periodic photonic crystals are already available in the form of photonic-crystal fibres, which use a nanoscale structure to confine light with radically different characteristics compared to conventional optical fibre for applications in non-linear devices, guiding exotic wavelengths, and so on.

Multilayer films (Bragg mirror) are perhaps the simplest forms of 1D photonic crystals and have been extensively studied by Lord Rayleigh since 1887. It was shown that any such 1D system exhibits a band gap for the transmission of electromagnetic waves. 1D-periodic systems are used in applications ranging from reflective coatings to distributed feedback (DFB) lasers. 2D-periodic optical structures were not studied in depth until the 1980s. The interest in two- and three-dimensionally periodic crystals with two- and three-dimensional band gaps is quite high in recent years (Fig. 4.5). The applications include LEDs, optical fibres, nanoscopic

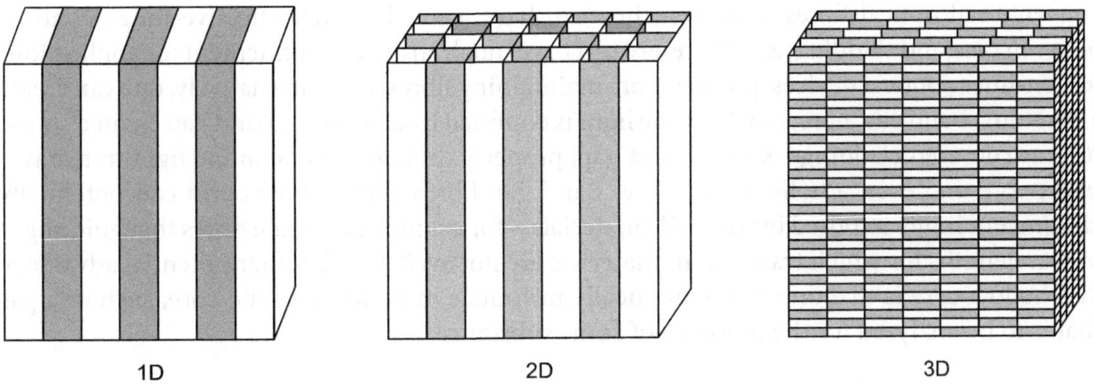

Fig. 4.5 Scheme describing different periodicity types in a photonic crystal.
(*Source:* http://commons.wikimedia.org/wiki/File:Dimensionphc.png).

lasers, ultra-white pigment, radio frequency antennas and reflectors, and photonic integrated circuits.

Photonic crystal fibre (PCF): This is a new class of optical fibre based on the properties of photonic crystals. Because of its ability to confine light in hollow cores or with confinement characteristics not possible in conventional optical fibres, PCF is now being used in optical communications, fibre lasers, non-linear devices, high-power transmission, highly sensitive gas sensors and other areas. More specific categories of PCFs include photonic-band gap fibre (PCFs that confine light by band gap effects), holey fibre (PCFs using air holes in their cross sections), hole-assisted fibre (PCFs guiding light by a conventional higher-index core modified by the presence of air holes), and Bragg fibre (photonic-band gap fibre formed by concentric rings of multilayer film).

In general, such fibres have a cross section (normally uniform along the fibre length) microstructured from two or more materials, most commonly arranged periodically over much of the cross section, usually as a 'cladding' surrounding a core (or several cores) where light is confined. Such fibres are constructed by the same general principles as other optical fibres: first, one constructs a 'preform' on the scale of centimetres, and then heats the preform and draws it down to a much smaller diameter (often nearly as small as a human hair), shrinking the preform cross section but (usually) maintaining the same features. In this way, kilometres of fibre can be produced from a single preform. Most photonic crystal fibres have been fabricated in silica glass, but other glasses have also been used to obtain particular optical properties (such as high optical non-linearity). There is also growing interest in making them from polymers, where a wide variety of structures have been explored, including graded index structures, ring-structured fibres and hollow core fibres. These polymer fibres have been termed MPOF or microstructured polymer optical fibres.

Photonic crystal fibres can be divided into two modes of operation, according to their mechanism for confinement. Those with a solid core, or a core with a higher average index than the microstructured cladding, can operate on the same index-guiding principle as conventional optical fibres; however, they can have a much higher effective-index contrast between core and cladding, and therefore can have much stronger confinement for applications in non-linear optical devices, polarization-maintaining fibres, etc. Alternatively, one can create a 'photonic band gap' fibre, in which the light is confined by a photonic band gap created by the microstructured cladding—such a band gap, properly designed, can confine light in a *lower-index* core and even a hollow (air) core. Band gap fibres with hollow cores can potentially circumvent limits imposed by available materials—for example, to create fibres that guide light in wavelengths for which transparent materials are not available. Another potential advantage of a hollow core is that one can dynamically introduce materials into the core, such as a gas that is to be analyzed for the presence of some substance.

4.2 MICRO- AND NANO-ELECTROMECHANICAL SYSTEMS (MEMS/NEMS)

Micro-electromechanical systems (MEMS) are being used in a wide range of engineering applications ranging from consumer products to defence systems, healthcare, automobile and environmental monitoring. Application of MEMS is motivated by several factors including their small size, reduced weight and power consumption, improved speed and precision, etc. Fabrication of MEMS necessitates the development of production technologies capable of precise dimensional control of fine features and those that are repeatable, reliable and not cost prohibitive. The dimensional features of MEMS vary in the range of few micrometres to few hundred micrometres. Microfabrication techniques are powerful tools for batch processing and miniaturization of electromechanical devices and systems into a dimensional scale which is not achievable by conventional machining techniques.

The electrical resistance of Si is a sensitive function of the external pressure and this effect is called the *piezoresistive* effect. This makes Si a good sensor material to detect pressure variations. Coupling silicon micromachining technology facilitates the fabrication of miniaturized micro-sensors. The magnitude of variation in resistance with change in pressure is dependent on several material and environmental factors like orientation of the crystal, type and concentration of impurity and temperature. MEMS-based silicon sensors can be used in the pressure range of 10^{-3} to 10^{6} torr.

NEMS (nano-electromechanical systems) based force sensors, chemical sensors, biological sensors and ultra-high frequency resonators, where structures of very small mass provide essential functionality, have also been fabricated. NEMS can be fabricated either by top-down or bottom-up processes. Presently, top-down methods using lithographic techniques are the most common. Bottom-up approaches involve the fabrication of nanoscale devices in much the same way as nature constructs objects—by sequential assembly using atomic and molecular building blocks.

Fabrication of NEMS structures is not limited to Si. In fact, III–V compounds such as gallium arsenide (GaAs) make particularly good NEMS materials from a fabrication perspective because thin epitaxial GaAs films can be grown on lattice-matched materials that can be used as sacrificial release layers.

One of the main constraints in the wide-scale application of MEMS/NEMS is expected to be in their packaging. Since each device will have a unique requirement for interacting/ protection from environment, it is difficult to standardize the packaging procedure. This lack of standardization tends to drive up the costs associated with packaging, making MEMS less competitive. In addition, it has to be ensured that the packaging mechanism does not interfere in the sensing mechanism or will not result in variations in sensor sensitivity during performance with time. Wafer-level packaging methods could hold the solution to this in the future.

4.3 NANOSENSORS

Nanosensors exhibit several distinct advantages over their micro scale and macroscale counterparts as listed below:

- Reduction in the overall size and weight of the associated system
- Cost reduction
- Mass production
- Utilization of physical phenomena appearing on the nanoscale
- Low power consumption
- Certain applications require nanoscale systems for functional applications, for example, implanted medical sensors must be in the nanoscale.
- Enhanced sensitivity
- Higher level of integration

Nanoscale sensors can be classified depending on the sensing application as physical, chemical or biological nanosensors. Also, similar to the classification of sensors, nanoscale sensors can also be classified according to the energy transduced (Table 4.1).

Other than these, nanosensors can also be classified according to:

- Effect/transduction phenomena
- Mcasurand
- Material of the sensor element
- Technological aspects

Table 4.1 Various nanosensors

S.No.	Type	Measured property
1	Mechanical	Size, velocity, acceleration, mass flow, force, torque, pressure, acoustic wave, piezoelectric, strain, stress
2	Thermal	Temperature, specific heat, entropy, heat flow, flux
3	Electrical	Voltage, current, resistance, impedance, inductance, capacitance, dielectric constant, polarization, electric field, frequency, dipole moment
4	Magnetic	Field strength, flux density, magnetic moment, permeability
5	Optical	Intensity, frequency, phase, wavelength, polarization, reflectance, transmittance, refractive index,
6	Chemical	Composition, concentration, reaction rate, pH, oxidation/ reduction potential

In particular, nanosensors can be classified depending on the nanostructures employed, such as, nanotubes, nanowires, nanoparticles, nanocomposites, quantum dots, embedded nanostructures, etc. Though many varieties of sensors exist, no single sensor can effectively sense all interested parameters in all possible environments. Therefore, building up of sensor arrays to consolidate multiple properties in different environments is of current interest. Sensor arrays contain different combinations of uni-and multi-functional sensors, to sense multiple phenomenon at one time, like the human sensor system with eyes as the optical sensor, nose as gas sensor, ear as the acoustic sensor and tongue as liquid chemical sensor. This increases data acquisition and multiplication and is used in the chemical or biochemical industries and so on.

Nanosensors and nano-enabled sensors have applications in many industries like transportation, communications, building and facilities, medicine, safety and national security, including both homeland defence and military operations. There are numerous examples of nanowire sensors that are used, for example, to detect chemicals and biologics: nanosensors are placed in blood cells to detect early radiation damage in astronauts, and nanoshells are used to detect and destroy tumours.

4.3.1 Carbon nanotube-based sensors

The electronic properties of carbon nanotubes are such that they may be metallic or semiconducting depending on their diameter and the arrangement of graphitic rings in the walls. They also exhibit exceptional mechanical, thermal and chemical properties. Utilization of these properties has led to the application of nanotubes as scanning probes, electron field emission sources, actuators and nano-electronic devices.

Their nanometric dimensions, high aspect ratio, large surface area and unique thermal, optical and electronic properties have promoted carbon nanotubes as one of the perfect candidates for sensing applications. Nanotube-based physical sensors can measure pressure, flow, temperature and the mass of an attached particle.

As discussed in Chapter 1, researchers at the Georgia Institute of Technology have demonstrated a carbon nanotube–based nanobalance which can weigh sub-micron scale particles. By applying an alternating voltage, they were able to create resonance in the nanotube with a specific frequency which depends on the nanotube length, diameter, density and elastic properties. The mass of the particle was calculated from changes in the resonance frequency that occur on placing the particle over the carbon nanotube. Using this technique, the mass of a carbon sphere was determined to be 22 femtogram, which is by far the smallest mass ever measured. This approach may lead to a technique for the weight measurement of individual biomolecules. This femto balance can also find application in weighing bio-organisms such as viruses.

Cleland and Roukes at the California Institute of Technology have reported the fabrication and characterization of a working nanometre-scale mechanical electrometer. The device has demonstrated a charge sensitivity below a single electron charge per unit bandwidth (~0.1 electrons/Hz at 2.61 MHz) better than that of state-of-the-art semiconductor devices and comparable with the charge detection capabilities of cryogenic single electron transistors.

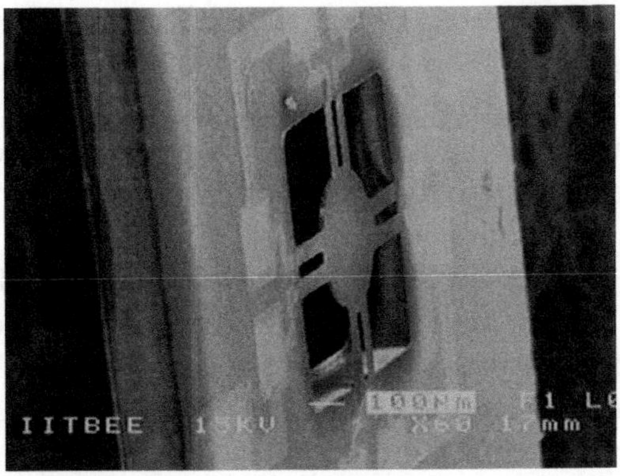

Fig. 4.6 MEMS accelerometer fabricated using polymer composite.
(*Source:* V Ramgopal Rao, IIT Bombay)

Prof. Ramgopal Rao and his group at IIT Bombay have developed a MEMS-based accelerometer using a polymer composite (Fig. 4.6), which can sense acceleration levels down to the 100 mg range. A low-cost technique at temperatures below 100°C has been used for the fabrication, which allows easy integration with CMOS.

Single-walled carbon nanotubes have been shown to exhibit piezoresistive effect, i.e. when they are bent or stretched, their electrical resistance changes. Based on this principle, carbon nanotube–based pressure and strain sensors have been developed. The pressure sensor consists of an ultrathin aluminium oxide membrane to which carbon nanotubes are attached. To calibrate the device, the deformation of the membrane in response to applied pressure was measured using white-light interferometry. They then monitored changes in nanotube resistance as a function of strain. They could detect a change even for strains as small as a hundredth of a percent, which in this case were induced by pressures of a few tens of kilopascals. The sensing nanotube was in this case metallic so that the gauge factor was positive. It had a value close to that of the best silicon devices.

Flow sensors have also been realised using SWNT. The SWNT bundles were packed between two metal electrodes and it was observed that they produced electrical signals in response to fluid flow. This is due to the direct scattering of the free carriers from the fluctuating coulombic fields of the ions or polar molecules in the flowing liquid. It was found through experiments that the ionic strength of the flowing liquid significantly affected the induced voltage.

The electronic properties, such as the local density of states, of single-walled carbon nanotubes are shown to be extremely sensitive to the chemical environment, because in this case all the tube atoms are surface atoms. The electrical conductivity and thermoelectric power also vary during exposure to several gas species and most of the nanotube-based chemical sensors are developed based on this principle.

It has been observed that the electrical resistance of individual semiconducting SWNTs changes by up to three orders of magnitude on exposure to NO_2 or NH_3

molecules at room temperature. Using this effect, a chemical sensor in which a single semiconducting SWNT was placed in contact with titanium or gold metal electrodes at the end has been fabricated. SWNT-based hydrogen sensors have also been demonstrated by sputter-coated individual and bundled SWNTs with Pd nanoparticles. The conductivity was found to decrease by 50% and 33% for the individual and nanotube bundles respectively, on exposure to an air mixture containing 400 ppm of hydrogen. Oxygen sensors based on change in resistance and thermoelectric power on exposure to oxygen have been reported. For biosensors, various ligands can be connected to a nanotube transducer in order to recognise target molecules specifically. The large surface areas available for molecule adsorption make carbon nanotubes a suitable material for biosensors. Carbon nanotube–based sensors are a thousand times smaller that microelectromechanical systems (MEMS) sensors and consume less power. These two important advantages will foster their use as implantable devices. Such biosensors based on functionalized carbon nanotubes will provide high sensitivity, large linear range, fast response, long life and low detection thresholds for different analytes.

4.3.2 Nanowire sensors

Nanowires are 1D structures a few nm in diameter and several microns in length and can be developed from a variety or combination of materials. They exhibit unique electrical and optical properties which can be exploited for sensing applications leading to ultrasensitive nanosensors. Sensors based on nanowires of various materials have been reported for biosensors, molecular detection, nano-connectors and gas sensors.

In general, the mechanism employed for inorganic nanowire sensors is field effect, which is transduced using FET. The conductance of a nanowire FET can be varied by the voltage of an applied gate. The molecules, and proteins, present on the surface of a nanowire influence the conductance significantly. Nanowire sensors modified with different antibody receptors can selectively recognize many different species in parallel and thereby enable high throughput screening for diagnostics and drug discovery.

FET sensors based on nanowires are more sensitive then planar FET sensors since binding to the surface of a nanowire leads to depletion or accumulation of carriers in the 'bulk' of the nanometre-diameter structure that increases the sensitivity to the point when detection of single molecules might be possible.

4.3.3 Polymeric nanofibres and nanocomposites

Sensors based on polymer structures have acquired great attention due to their unique advantages. Conductive polymer-based sensors exhibit higher sensitivity and shorter response time at room temperature over conventional sensors (usually made of metal oxides). Ease of synthesis (by chemical and electrochemical techniques) and modification of the main chain (by copolymerization and structural deviation) have resulted in enhanced application of conductive polymers for sensor applications. Nanofibres and nanocomposites are the current advanced polymer structures used for sensor applications.

Reduction in the diameter of polymer fibres to the nanoscale facilitates extremely good properties due to increase in surface area to volume ratio. Superior mechanical properties like stiffness and tensile strength compared to any other known form of material and flexibility in surface functionality make polymer fibres potential candidate for many applications. Polymer fibres can be coated with various synthesising techniques like electrospinning, template synthesis, self-assembly and phase separation, etc. Though all these techniques can be used to synthesize nano objects with complex architectures, electrospinning is the most preferred for polymer nanofibres.

The electrostatic spinning technique was developed in the 1930s and it involves an electro-static field to pull the fibres from the polymer solution. Electrospinning mainly consists of a high-voltage power supply, a capillary tube with a needle having a small diameter and a metal collector. High voltage supply to the electrodes with opposite polarity induces charge at the capillary tube end containing polymer solution held by its surface tension force. Increasing intensity of the electric field forms a Taylor cone (conical shape) at the capillary tube end and further increase in the field causes ejection of the fluid from the capillary tube. The instability in the discharged polymer solution causes elongation in the jet to make it thin and long. Evaporation of the solvent results in the formation of a nanofibre of polymer. Figure 4.7 shows a polymer nanofibre made by the electrospinning process.

Though the process is very simple, the preparation of polymer solution poses difficulties while dissolving polymers in the solvent and this process also needs high DC voltages up to 10 kV. Even polymers dissolved in various solvents can be electro-spun. In case of polymers with a high melting temperature, molten polymer itself can be introduced into the capillary tube instead of the polymer solution, provided the whole setup is located inside a vacuum chamber. Different nanocomposites have also been prepared by electrospinning of the

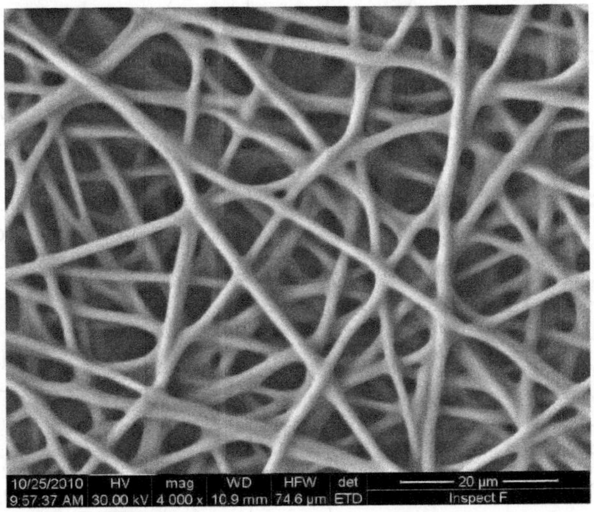

Fig. 4.7 Polymer nanofibres synthesized by electrospinning technique.
(*Source:* TS Sampath Kumar, IIT Madras)

solution mixture consisting of nanoscale organic or inorganic fillers added to the solvent. The alignment of single-walled carbon nanotubes (SWCNTs) was successfully achieved by electrospinning and used as reinforcement for the polymer composites to achieve good properties. Figure 4.8 shows the various applications of the polymer fibres prepared by electrospinning.

Nanocomposites encompass a large variety of systems composed of dissimilar components that are mixed at the nanometre scale. The behaviour of nanocomposites is dependent not only on the properties of the components, but also on morphology and interactions between the individual components, which can give rise to novel properties not exhibited by the parent materials. The size reduction from micro-composie to nanocomposites yields an increase in surface area which is utilised in sensing applications.

Polymer-based nanocomposites have been prepared by adding various inorganic compounds in order to achieve better properties like modulus, strength, heat resistance and impact resistance with decrease in dielectric constant and electrical conductivity. Reinforcement of nanocomposites is the main application of nanofibres. Nanoscale filler dispersed polymer nanocomposites with extraordinary properties have made them an alternative for traditional polymers. In literature, authors observed an increase in tensile strength, chemical resistance and heat resistance in layered silicate/polymer nanocomposites. Ferrites dispersed polymer based magnetic materials are largely replacing traditional ones due to their lower cost, and these materials are have great potential as microwave absorbers, sensors and in aerospace applications. Recently, in colour printers and copiers, γ-Fe_2O_3/polymer nanocomposites are being used as toner material and these are also used as contrast agents in NMR analysis and

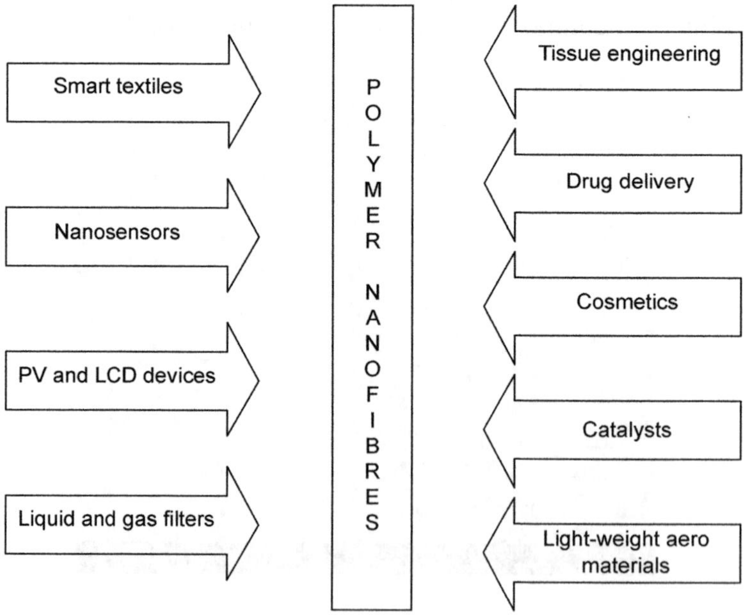

Fig. 4.8 Applications of electrospun ultra-nanofibres.

memory devices. Polymer core covered with magnetic materials are used as beads for gas separation, pigments, catalysts, toners, flocculants, coatings for anticorrosion protection and drug delivery.

Gases with an acid base or oxidising characteristics can be accurately sensed by conductive polymers. Polymers with active functional groups and solid polymer electrolytes (SPEs) are also used to detect such gases. Opto-chemical detection of HCl has been demonstrated using mono substituted tetraphenylporphin polymer thin films, with a sensitivity up to sub ppm level HCl using a polyhexylmethacrylate matrix.

4.3.4 Nanoparticles

Nanoparticles are clusters of atoms with sizes similar to the de Broglie wavelength associated with valence electron. So they can be considered as quantum dots having size-dependent discrete energy levels. Noble metal nanoparticle sensors work based on the high sensitivity of the LSPR (localised surface plasmon resonance) spectrum towards change in the local refractive index of nanoparticles when a foreign atom attaches to them. LSPR-based sensors are gaining more attention as biosensors. Magnetic nanoparticles are also used for biomedical applications by enriching the analytes to be detected with magnetic nanoparticles, either as ferrites like $MO.Fe_2O_3$ (M= Ni, Zn, Mg, Mn and CO) or in the form of superparamagnetic magnetite (Fe_3O_4), greigite (Fe_3S_4) and maghemite (γ-Fe_2O_3).

Particle size–dependent properties of noble metal nanoparticles and semiconductor quantum dots have made them potential candidates for optical nanosensors. Quantum dots (QDs) of nanocrystalline inorganic semiconductor materials are used as optical sensors based on their fluorescence measurements. In these semiconductor nanocrystals, the band gap can be tuned with the crystallite size; for instance, the smaller the size, the wider the energy gap and the shorter the wave length of fluorescence. Therefore, all fluorescence colours in the visible region can be obtained by optimizing the crystallite size during synthesis. QDs functionalised with antibodies could be used for environmental purpose like detection of pathogens (cholera toxin or ricin) in water. Intense absorption of visible/UV region observed in smaller nanoparticles (diameters less than the de Broglie wavelength of electron) of noble metals is unusual in bulk materials.

The use of chemical sensors for environment monitoring is of great importance. Nanoparticles are also very well suited for chemical sensor applications because their large surface area to volume ratio and physical properties often vary considerably in response to changes in the chemical environment. Because nanoparticles can be made of structures just a few atoms across, even a few molecules can trigger a response.

4.3.5 Plasmonic-based nanoprobes

The term plasmonics is derived from 'plasmons', which are the quanta associated with longitudinal waves propagated in matter through the collective motion of large numbers of electrons. When light is shined on there surfaces, conduction electrons are excited causing excitation of surface plasmons. This can lead to electromagnetic enhancement for ultrasensitive detection such as surface-enhanced Raman scattering (SERS) and surface-enhanced fluorescence (SEF).

Raman spectroscopy is widely used as an analytical tool. This is partly due to its non-destructive nature and structural fingerprinting capability with very narrow and highly resolved bands (0.1 nm). The other advantages are rapid spectral measurement and easy sample preparation. Low sensitivity and the requirement of powerful and expensive lasers for excitation are some of the drawbacks of conventional Raman spectroscopy. It has been shown earlier in the 1970s that the Raman signal is enhanced when molecules were adsorbed onto specific substrates, and this effect is known as surface enhanced Raman scattering (SERS) spectroscopy. The localized fields due to surface plasmon resonance and chemical effects were found to cause SERS enhancement. Silver-coated active nanospheres are used for sensitive detection of a variety of compounds of environmental and medical interest. SERS has also been used for gene probing, wherein selective detection of HIV DNA and the cancer gene was demonstrated.

Due to their non-radioactive nature, there is strong interest in the development of optical techniques for biomedical diagnostics, pathogen detection, gene identification, gene mapping and DNA sequencing. One of the examples of molecular recognition events is the hybridization of a nucleic acid to its target. This can lead to highly accurate complementary DNA sequences.

4.3.6 Optical nanosensors

Fibre-optic sensors provide useful tools for remote in situ monitoring. Fibre-optic sensors can be used for sensing intracellular/intercellular physiological and biological species. Scientists have developed nanosensors for the in situ intracellular measurement of single cells using antibody-based nanoprobes. Incubation of cells with fluorescent dyes or nanoparticles and their interaction is commonly studied using microscopy techniques. During incubation of a dye or nanoparticle into a cell, it is generally transported to intracellular sites. With optical nanosensors, excitation light can be delivered to specific locations inside cells. In these nanosensors, the monitoring process is almost non-invasive. Combined with the exquisite molecular recognition of antibody probes, nanosensors could serve as powerful tools for exploring biomolecular processes in sub-compartments of living cells. They have a great potential to provide the necessary tools to investigate the multi-protein molecular machines of complex living systems and the complex network that controls the assembly and operation of these machines in a living cell. It is expected that nanosensors equipped with nanotool sets will soon be developed for tracking, assembly and disassembly of multi-protein molecular machines. Scientists have so far been investigating the genes and proteins by breaking the cell apart and studying their individual components in vitro. It is expected that nanosensors would permit research on entire networks of genes and proteins in the living cell in vivo.

4.3.7 SQUID-based magnetic nanosensors

Semiconductor quantum interference devices (SQUID) were first invented in 1964 which involves obtaining small changes in current and magnetic fields as voltage at room temperature by using semiconductivity (Josephson effect and quantum interference). To avoid the interference of the ambient magnetic field, a separate pick-up loop (manually wound superconducting material wire) is used and is connected to an input loop. Pick-up will have some inductance (L_p) which is different from the inductance of both the input loop (L_i) and SQUID (L). Both the input loop and SQUID used to be shielded from the ambient magnetic fields with a niobium canister.

This device produces measurable voltage corresponding to the magnetic field. Among all the techniques available for measuring the magnetic field, SQUID is the most sensitive. These are used in magnetocardiology, non-destructive evaluation, geophysics and explosive detection. SQUID devices can be generally classified as radio frequency SQUIDS (semiconductor loop interrupted with one Josephson junction) and direct current SQUIDS (semiconductor loop interrupted with two Josephson junctions). However, in many applications, Dc SQUIDs are more advantageous than Rf SQUIDs. SQUIDs can also be classified as low-temperature SQUIDs (LTS) and high-temperature SQUIDs (HTS) based on their operating temperatures.

High-temperatures SQUIDs were invented in 1986, allowing the operation of SQUIDS at higher temperatures (operates at liquid nitrogen temperatures 77 K) whereas LTS operate at liquid helium temperatures (4.2 K). LTS are mostly fabricated from metallic superconducting materials and are generally isotropic with coherent lengths of few tens/hundreds of inter-atomic distances. HTS are generally ceramic, anisotropic and fragile with small coherent lengths. The isotropic nature and long coherent lengths of LTS allows them to form complex 3D and multi-layer structures such as axial gradiometers and enables them to operate NbTi-based LTS in high fields. In case of HTS, it is not possible to make axial gradiometers because of the smaller coherent lengths. Another advantage of LTS is that they are stable in air whereas HTS will degrade with moisture. Even though HTS are not as sensitive as LTS, the advantage of operating at liquid nitrogen temperatures made them obvious candidates due to easy availability of liquid nitrogen when compared to liquid helium. The ability of LTS to form complex 3D structures allows the detection circuit to reject the significant effect of the external noise which leads to better sensitivity of LTS. Other advantage of Nb-based LTS are that they are capable of operating in magnetic fields of tesla range which allows the LTS microscope to measure susceptibility. The most commonly used SQUIDs are made up of niobium (for LTs) and $YBa_2Cu_3O_{7-x}$ (for HTs).

Since current is always associated with a magnetic field, all physiological organs are associated with some magnetic field which enables SQUID to be used for the diagnosis of biological organs like the heart, brain, lungs and muscles. Tiny magnetic fields generated by the nerve current while controlling the physiological activities are monitored to identify abnormal conditions due to diseases. These magnetic fields can range from a few fT (response from brain) to a few nT (lung response). This makes SQUID the only technique with the capability to measure it. Particularly low-temperature SQUID-based sensors are used for brain imaging. The low response time of SQUID enables time-resolved studies on brain activity. SQUIDs are also evolving as potential candidates for magneto-encephalography (MEG), magneto-cardiogram (MCG) and for monitoring neural disorders.

The earth's magnetic fields at frequencies ranging from 1 kHz to 10^{-4} Hz can be detected with SQUIDs and can be classified as seismic (acoustic energy), gravity, electromagnetic, radiometric, electric, etc., based on the method used to determine the geophysical magnetic field. The electrical and electromagnetic methods are more sensitive towards the resistivity of the subsurface materials in frequency ranges below 100 kHz. SQUIDs are sensitive to both resistance and dielectric constant of subsurface materials above 100 kHz.

The skin depth limitation of eddy current testing means that SQUID sensors can be used for the analysis of flaws in metallic materials due to its superior sensitivity and DC response which allows higher depths. DC response of the SQUID sensors can measure remnant magnetization without application of external magnetic field. Flaws and perturbations can be detected irrespective of the magnetic field. SQUID sensors use the advantage of magnetic sensing (i.e., the insulator covering of the materials to be tested will be invisible for magnetic detection) to extend its applications in the area of stress and corrosion detection in reinforced rods used in big constructions and aircraft runways. An increase in the number of ageing aircraft would result in structural failure due to mechanical effects or hidden corrosion effects. Addressing the problem with great sensitivity makes SQUID-based systems a powerful tool for non-destructive testing. These sensors have also been used as aircraft wheel testers to detect defects in wheels.

The short circuit in ICs appears as intense magnetic flux in a small region which could be plotted by magnetic mapping using SQUID sensors. The advantage of SQUID-based magnetometers is good resolution, and magnetic mapping can be done in the non-contact mode without destroying the electric circuit.

4.3.8 Biosensors

Nanosensors are used in medicine in diagnosis and in aiding controlled drug delivery to infected cells. Several biological nanosensors for cancer detection, blood glucose detection, biotoxin detection, detection of specific DNA and identification of chromosomal defects have been realised already.

Microcantilever-based cancer detection tools have also been developed. The microcantilevers are coated with appropriate antibodies that can bind the prostate specific antigen (PSA) in cancer containing blood. The antigen–antibody adhesion will result in change in mass and hence bending of the cantilever which can be optically detected. Such microcantilever-based cancer detectors have been found to be more sensitive them other conventional biochemical techniques available. Similar cancer detectors using resonant frequency shift of piezoelectric nanomechanical microcantilever have also been demonstrated.

Using a similar principle of coating specific materials over microcantilevers, myoglobin and glucose sensors have also been developed. Microcantilever-based biochips for detecting biotoxins like anthrax and aflatoxin have been developed and can find possible application in providing security against biological warfare and terrorist attacks. The gold side of a biomaterial microcantilever (Au–Si) is coated with a receptor that binds to specific biological molecules.

4.3.9 Microcantilever-based sensors

Microcantilever-based sensors can be used for a wide range of applications ranging from detection of chemicals to surface stresses. With advancement in the fabrication technologies to produce microcantilevers in large numbers and at low cost, their commercial application is gaining renewed interest in the field of chemical, physical and biological sensor development. External environmental factors like adsorbed chemicals or stress can lead to change in the resonance frequency, amplitude and deflection response of the microcantilever. Microcantilever deflection can be measured by any of the following techniques: piezoresistive deflection detection method, optical deflection detection method, capacitive deflection detection method, interferometry deflection detection method, optical diffraction grating deflection detection method or by the charge-coupled device (CCD) detection method.

Sensors based on microcantilevers primarily rely on detecting the changes in bending or vibrational frequency to act as a physical, chemical or biological sensor. Silicon, silicon nitride or silicon oxide are commonly used as commercial cantilevers. Cantilever arrays have also been fabricated recently employing the latest integrated circuit (IC) and complementary metal oxide semiconductor (CMOS) technologies.

It is possible to coat a thin layer of specific material over microcantilevers. These coatings are chosen to provide selective adsorption capabilities to certain molecules. When these molecules are adsorbed on a microcantilever, they change the vibrational frequency and the deflection of the microcantilever. The magnitude of variation will provide an estimate of the adsorbed mass. It has been demonstrated that microcantilever-based mass detection sensors can measure a change of about 0.3 attograms (1 attogram = 10–18 g). When the cantilevers are coated with a certain receptor that can have specific binding to desired biomolecules like proteins, it is possible to extend these sensors for biomedical application. Microcantilever-based sensors are being evaluated for screening diseases such as cancer and detecting specific chemical and biological warfare agents.

The advantage of using microcantilever-based technology for biosensing applications is that they are fast, easy-to-use, cheap and highly sensitive for detecting analytes. Usually, measurement of trace elements requires sophisticated technologies like high-performance liquid chromatography (HPLC), thin layer chromatography (TLC), gas chromatography (GC), and gas liquid chromatography (GLC). In comparison, microcantilevers are less costly, consume less power and space and have the advantage that they can detect even trace quantities in ppb and parts per trillion concentration.

Diverse fields like biomedical analysis, cell incubation and environmental monitoring require highly sensitive humidity sensors. It has been noted that there are various humidity sensors like hygrometric, gravimetric, capacitive and optical humidity sensors available. However, microcantilever humidity sensors find great interest where humidity measurement is enabled by coating one side of the cantilever with materials like gelatin that have high binding tendency with water vapour in the atmosphere. Deflection of the cantilever due to adsorption of water vapour is measured and calibrated as a measure of the humidity in the environment.

By employing bimetallic coatings over microcantilevers, it is possible to detect changes in temperature and heat bend as a function of cantilever bending due to differential thermal expansion coefficient. The extent of bending will increase with increase in temperature. It is also possible to coat the cantilever with IR absorbing coatings to result in remote IR detection capabilities. Temperature variations as small as 10^{-5} K have been detected using this technology. Thus, these sensors can also find applications in microcalorimeters.

The vibration amplitude and frequency of the microcantilever is also a function of the damping capacity of the environment. A viscous medium will lower the resonance frequency, enabling the detection of viscosity variations using microcantilevers. It is possible to use piezoelectric actuators to resonate the cantilevers for use as viscosity meters.

Micromechanical sensors coated with different materials can be used to detect electromagnetic and nuclear radiation. When micromechanical sensors absorb radiation, they bend and their resonance characteristics change. Optical radiation dosimeters are formed by coating microcantilevers with optical cross-linking polymers. Similarly, cantilevers made of materials which are sensitive to various nuclear particles or radiation can be used for nuclear radiation dosimetry.

4.3.10 Electronic nose

A rapid sensory information and low-cost electronic instrument which can mimic the human olfactory process is of great interest in food processing, volatile gas analysis, forensics and environment monitoring. This electronic instrument mimics the human biological sensor system by exhibiting a unique response to each odourant.

The electronic nose consists of a sample holder, an array of non-specific electrochemical sensors and data collectors or signal processing in series. The array of chemical gas sensors with a unique response profile or different sensitivity for each individual odour, records the profile of all the sensors in the array used to characterize odour. The output from the electronic nose gives the characteristics and concentration of the odour similar to the human sensing system.

The analysis of an odour with the electronic nose involves the absorption of gas with an odour by a vacuum pump which allows it to pass through an array of sensors. The molecules with an odour are adsorbed by the sensors, generate a response profile that is recorded and sent to signal processing unit. Finally a carrier gas is pumped over the array of sensors to make them ready for the next analysis. Most commonly used sensor arrays for the electronic nose are made of conducting polymers or metal oxide sensors; both give resistance when exposed to volatile compounds. Commonly used and commercially available metal oxide semiconductor sensors are based on oxides of tin, zinc, titanium, tungsten and iridium doped with noble metal catalyst of platinum or palladium. This involves the deposition of the volatile organic compounds on the semiconductor sensors in contact with two metals at 200 to 400°C. The adsorption of the volatile compounds alters the resistance of the metals which is the indication for identification of volatile compounds. The second category of sensors is made of conductive polymers; the most commonly used is polypyrroles, thiophenes, indoles and furans. The bonding of the volatile molecules to the polymer chains during exposure to various chemical results changes the conductivity of these materials. The electronic nose has been used to analyse the gas from spoilage microorganisms of milk and food items. The main advantage of the electronic nose is its portability, user friendly operation, spead and reliability compared to gas chromatography and mass spectroscopy (latter used to take 20 to 100 minutes to analyse once whereas the former takes a few seconds to minutes).

4.3.11 Electronic tongue

An electronic tongue is a multisensor system with an array of chemical sensors and a pattern recognition device. In Japan, the electronic tongue is also known as the 'taste sensor' because of its analogy with the human sensor system. This electronic instrument is used for liquid analysis and is similar to the electronic nose in analysis sequence but the latter is used for chemical gas analysis. Various measurement techniques like potentiometric, voltammetric, conductometric and spectrophotometric are commonly used in electronic tongue for analysis. In potentiometric sensors, electronic tongues contain either arrays of glass electrodes or a polymer membrane (vinyl chloride). Change in charges of these gives an indication of ions or the taste of the beverages like tea, coffee and milk. A photometric electronic tongue contains a charge-coupled device with a chemical indicator in a resin bead used for the analysis of simple sugars at different pH values. Conductometric electronic tongues measure conductivity to give the fat content of milk. Voltammetric tongues measure current at constant voltages. This type has been used extensively because of its high sensitivity, simplicity and robustness. It can be used to classify fruit juices, milk, etc. Though the electronic tongue was initially used only for the analysis of liquids, it can also be used for liquid–solid mixtures and homogenates.

4.4 NANOCATALYSTS

Chemical reactions are significantly enhanced by catalysts. Due to the large surface area-to-volume ratio, nanomaterials can be more efficient catalysts. Nanocatalysts can lead to cost savings and can also have higher selectivity than conventional catalysts; this can reduce waste and hence the environmental impact.

A catalyst is a substance that changes the rate of reaction without itself being consumed in the reaction. When we say it is not consumed, it does not mean that the catalysts are not active participants in the reaction; they are active. The catalyst usually reacts with the reactants to form a stable complex:

$$\text{Reactants} + \text{Catalyst} \longrightarrow \text{Complex}$$

The complex rearranges to yield the products and regenerates the catalyst:

$$\text{Complex} \longrightarrow \text{Products} + \text{Catalyst}$$

Notice that the catalyst is regenerated at the end of the reaction, so there is no net consumption of catalyst. A catalyst is considered to be active in any given chemical process if it shows high conversion, is selective to the desired products, stable for a prolonged period of time and has good mechanical strength. Of all the concerns, conversion and selectivity dictate the fate of a catalyst, as they can significantly change the economics of the process. In general, high conversion can be achieved if the catalyst species is not sintered during the reaction, and selectivity is achieved from the specific crystal structure of the catalytically active metal or metal oxide precursor. Hence, controlling the catalyst species at a molecular level is possible if the catalysts are fabricated at the nanoscale. Particles of nano-size have definite crystal structure, and hence the application of nanostructured materials as catalysts can drastically change the conversion and selectivity in the chemical processes.

Particles in the 1–100 nm range are opening up new vistas in surface chemistry, for two reasons:

- **Huge surface area** with many of the atoms on the surface, thus allowing good 'atom economy' in surface–gas, surface–liquid and surface–solid reactions.
- **Enhanced reactivity** due to crystal shape. For example, the shape changes from cubic to polyhedral, when the surface concentration of edges and corner sites goes up considerably.

However, there are other features that can affect 'surface energy'. As the crystal size becomes smaller, anion/cation vacancies of the Frenkel or Schottky type become prevalent. Also, atoms on the surface can be distorted in their bonding patterns. Nanostructured catalysts can be divided into two classes:

- Catalysts composed of nanoparticles supported on metal oxides or molecular sieves
- High surface area nanocrystalline metal oxides as catalysts or catalysts supports

Since, the catalytic sites of metal catalysts are located on their surface, nanoparticles with enhanced surface-to-volume ratio are expected to be very effective. Figure 4.9 shows the dependence of the fraction of surface atoms on the radius of nanoparticles. The figure shows that the smaller the size, the larger the ratio. Another characteristic property of nanoparticles is the quantum size. Although the bulk metal has a band structure, the

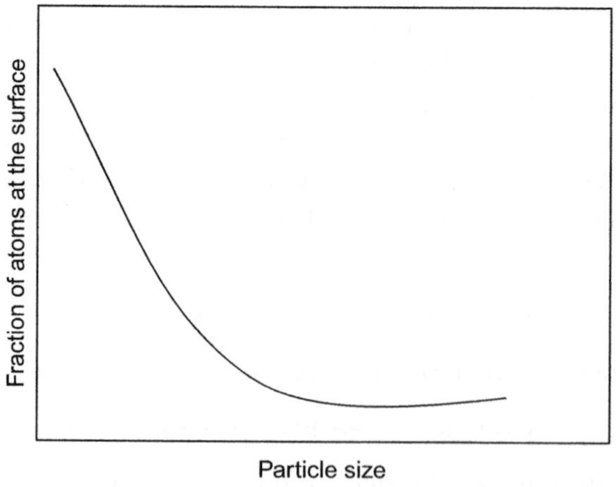

Fig. 4.9 Schematic showing the influence of particle size on the fraction of atoms at the surface.

electronic energy levels of metal nanoparticles—with size of a few nanometres—are rather separated.

Metal nanoparticles with rather uniform size can be prepared by both physical and chemical methods. In the physical method, or the top-down method, metal nanoparticles are prepared by decomposition of bulk metal using mechanical force, vaporisation, laser abrasion, and so on, which can provide higher energy to bulk metal than the bond energy of the metal. In the chemical method, or the bottom-up method, the preparation process starts from reduction of metal ions to metal atoms followed by aggregation, resulting in metal nanoparticles. Both methods have their own advantages. However, chemical methods are considered to be better from the viewpoint of reproducibility, homogeneity and mass production.

4.4.1 Gold nanoparticles

Gold nanoparticles have attracted attention as catalysts because of their wide application in CO oxidation, epoxidation of propylene, water gas shift reaction, hydrogenation of unsaturated hydrocarbons and liquid phase selective oxidation. It is well known that gold is a poor catalyst in the bulk form, whereas gold nanoparticles supported on Fe_2O_3, Co_3O_4 and NiO showed high catalytic activity for low temperature oxidation of carbon monoxide. CO oxidation is very important in the purification of engine exhaust gases, hydrogen produced by steam, reforming of methanal and hydrocarbons for polymer electrolyte fuel cells.

It has also been shown that in an Au/TiO_2 catalyst, the change in turnover frequency (TOF) of CO oxidation changes with the diameter of Au islands. Some have suggested that this transition might be correlated to the high catalytic activity. Hayashi and co-workers found that Au supported on TiO_2 could catalyse epoxidation of propylene in the gas phase containing O_2 and H_2. There are several factors that influence the catalytic activity of the reaction. The most important is the method of preparation. Au/TiO_2 was prepared by either an impregnation method or a deposition precipitation method. The impregnation method did not result in selective oxidation, but instead in complete oxidation to H_2O and CO_2; whereas the direct precipitation method led to epoxidation with selectivity above 90%. The difference in the structure of the catalysts prepared by different methods influences the selectivity. The impregnation method produces large, spherical Au particles with diameter of about several tens of nanometres, whereas, the direct precipitation method results in small hemispherical Au particles, strongly contacted with TiO_2 support. This is a very interesting example of how the structure and size of nanoparticles affects the selectivity of the reaction.

4.4.2 Magnetic nanoparticles

Magnetic nanoparticles are being used for a variety of applications such as catalysis, drug delivery, etc. Silica and carbon are used to maintain the stability of these nanoparticles. The nanoparticle surfaces are functionalised for catalytic activity with species such as enzymes. The magnetic nanoparticle catalysts can be easily separated by an external magnetic field.

4.4.3 Other nanocatalysts

Nanopowder catalysts with silica and platinum nanoparticles exhibit very strong catalytic activity for hydrolysation reactions. Carbon–carbon coupling reactions occur in a wide variety of solvents using dendrimer-encapsulated metal nanoparticles. Titania-based nanocatalysts are being increasingly used in photocatalysis. Photocatalytic reactions are of considerable interest because of their applicability in the treatment of pollutants and wastes and in the utilization of solar energy. TiO_2 has high oxidising power that is capable of oxidising organic carbonaceous substances to CO_2 in the presence of water and oxygen. Nanocrystalline alkaline earth metal oxides have attracted considerable attention as effective absorbents for toxic substances such as NO_2, SO_2 and HCl. Nanocrystalline MgO particles act as an effective catalyst for dehydrogenation.

4.5 FOOD AND AGRICULTURE INDUSTRY

Some of the possible areas of application of nanoscience in agriculture and food processing include:

- Nano-porous zeolites for slow and efficient release of water and fertilisers for plants, and nutrients and drugs for livestock
- Nanocapsules for herbicide delivery
- Nanosensors for soil quality and plant health monitoring
- Nanocomposites for plastic film coatings used in food packaging
- Antimicrobial nano-emulsions in the decontamination of food equipment, packaging or food processing
- Nanoscale biosensors for the detection and diagnosis of pathogens. A nano-bioluminescent spray that reacts to pathogens by producing a visual glow has been developed. The product can be applied on the contents of refrigerated freight containers to detect bacteria quickly and economically. Another application for the spray might be to detect bioterrorism. This can have a huge impact on food security and enhance food quality for all.
- Development of nanobarcodes as identification tags for food and agricultural products.
- Food supplements, colour additives and animal feeds

The use of nanosensors in food packaging and agricultural products can ensure that food is free from contamination and provide safeguards against bioterrorism. Photocatalytic degradation for wastewater treatment and as a disinfectant, particularly in fruit packaging, is currently being studied. Nanocrystalline oxides of metals like Ti, Sn and Zn have been found to be good photocatalytic materials in aiding disinfection of packed food and agricultural products. The high surface area-to-volume ratio also becomes advantageous in the use of nanomaterials for these purposes, compared to conventional materials. The photocatalytic process releases excited electrons. These electrons can enter the bacteria adhering to the nanoparticles, thereby resulting in disinfection.

Nanoparticles capable of biofluorescent recognition are being developed as beneficial substitutes for organic dyes currently being used. The luminescence has been found to be more efficient for fluorescent labelling using quantum than with organic dyes, since their emission spectra are narrow, symmetric and tuneable, according to the particle size and material composition of the QDs; they also exhibit good photostability.

Another interesting application is in the area of *nano-modified seeds* and fertilisers/pesticides. Atomic engineering can modify the DNA of seeds to obtain the desired yield. Nano-modified food products are being studied to enhance the nutritional content of the food, by using nano-encapsulated nutrients, disabling fat and sugar molecules using nano-modification, and adding taste boosters. This can make a huge difference to the fast food industry—food can be less unhealthy and more attractive to customers. A major benefit of nanotechnology is likely to be in the food packaging industry, where it can be used to increase the shelf life of food products and for many other applications, as shown in Fig. 4.10.

4.6 COSMETICS AND CONSUMER GOODS

Nanomaterial-based cosmetic products are becoming increasingly popular. Currently, more than a few hundred nanomaterial-based cosmetics, sunscreens, anti-ageing and personal care products are commercially available.

4.6.1 Sunscreens

Sunscreens containing titanium dioxide or zinc oxide particles are commonly used to prevent skin burn, since these materials have a good absorption coefficient for UV rays in

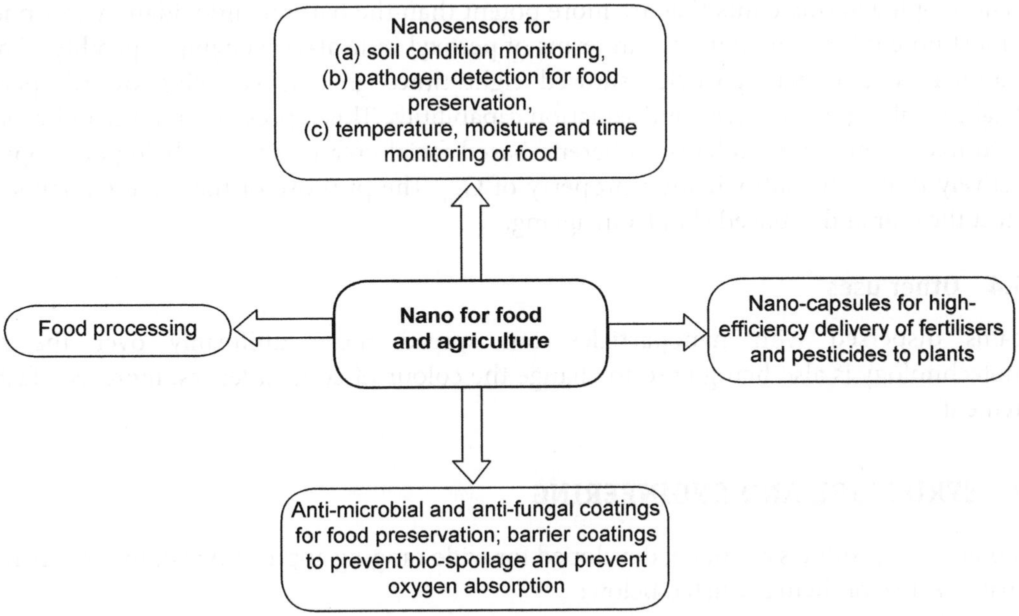

Fig. 4.10 Applications of nanotechnology in the agriculture and food sectors.

particular. These creams (based on microcrystalline particles) are usually white in colour and are distinctly seen on application over the skin. However, the use of nanocrystalline titanium dioxide and zinc oxide particles in these creams makes them transparent, resulting in the skin being clear and visible even after application of the cream. The physical appearance is not hampered by the application of these transparent creams. Hence, these nanoparticle-based sunscreen creams are becoming increasingly popular. The use of nanoparticles also results in (i) good spreading behaviour of the cream and (ii) more effective absorbtion, particularly in the ultraviolet spectrum of light.

4.6.2 Personal care products

The use of nanomaterials imparts anti-bacterial and anti-odour functionality to powder, gel, stick and spray underarm products. They also make them glide smoothly on the skin. Nano-gold and nano-silver particles have anti-bacterial and anti-fungal characteristics. They are hence used in many personal care products. Silver particles of size less than 10 nm dispersed in soaps impart the double advantage of killing germs and increasing effectiveness in removing dirt particles from the skin. Deodorants and antiperspirants based on nano-zinc oxide particles are currently being developed. Silver-doped zinc oxide offers enhancement in odour fighting and anti-bacterial properties.

4.6.3 Anti-ageing products

Traditionally, vitamin E has been regarded as an antioxidant against peroxyl radicals and thereby reduces the ageing kinetics of biological cells. For decades, considerable effort has gone into developing antioxidants that are more potent than the natural antioxidant α-tocopherol, the most potent form of vitamin E in terms of peroxyl radical–scavenging capability. Trolox-functionalised gold nanoparticles showed eight times greater reactivity towards peroxyl radicals, resulting in enhanced antioxidation capabilities. This represents a novel and versatile way to make better antioxidants. Fullerene-based facial creams are also being developed to effectively utilise the antioxidation property of C_{60}. The purpose of these antioxidants is to protect the hair and exposed skin from ageing.

4.6.4 Other uses

Creams dispersed with nanoparticles can spread more uniformly over the skin. Nanotechnology is also being used to change the colour of contact lenses, more as a fashion statement.

4.7 STRUCTURE AND ENGINEERING

Nano-enabled products are being developed for wide ranging applications in the construction industry. A few of them are listed below:

- Reinforcement of concrete with nanoparticles
- Self-cleaning glass

- Smog-eating concrete
- Nano-enabled wood
- Intelligent sensors to enhance the safety of buildings against natural and other calamities (sensors to sniff toxic substances)
- Fire-resistant coatings
- UV/IR reflecting windows

The use of *nano-silica* in concrete has been reported to increase its mechanical strength as it results in better packing density. Also, it decreases the kinetics of chemical degradation of concrete by calcium silicate hydrate reaction and decreases water penetration in concrete, leading to better life and durability. Concrete reinforced with nano-TiO_2 or carbon nanotubes has been found to have remarkable increase in strength and wear resistance. Similarly, nanocomposite steel has been reported to have at least three times higher strength than conventional steel.

Impregnation of nano-ZnO, -Ag or -TiO_2 particles in wood can help to reduce their biodeterioration. This could be a potential substitute for the current practice of using toxic chemical treatment to prevent biodeterioration of wood. Wood itself is made of cellulose nanofibrils and has about one-fourth the strength of the carbon nanotubes. Development of a technology to extract these nanofibrils from wood can result in an economic substitute for hard composite-based construction material, since they are expected to be 10–100 times cheaper than carbon nanotubes. Novel fire retardant nanocrystalline coatings are also being developed with potential applications in constructions. Most fire retardant coatings work by suppressing flames through intumescences, i.e., they puff up on exposure to flames or excessive heat, solidifying into foam. This foam insulates the substrate from the flame.

One of the most important emerging applications of nanotechnology in civil engineering could be the integration of sensors to monitor the integrity and safety of the construction. This could result in early assessment of damage after a calamity or even after terrorist sabotage. Sensors can also be used to sniff toxic gases and substances and for the early detection of wood destroying termites and fungi.

4.8 AUTOMOTIVE INDUSTY

The next generation automobiles aim to be lighter, more fuel efficient and crash resistant. They use high strength-to-weight ratio frames, use eco-friendly energy sources, pollution free emissions, producer sensors and functional nanomaterials for enhanced safety and aesthetics. Nanotechnology can play a major role in realising many of these objectives enhance vehicle performance, convenience and safety. A few of these applications are listed below:

- Carbon nanotube-based composites are being examined as a replacement for automobile frames due to their high strength and reduced weight. The unique combination of stiffness and toughness of nanopolymer composites will make them ideal substitutes for steel in automobiles. Also, clay nanoparticles in composite materials are being evaluated in cars. The use of materials with high strength-to-weight ratio for body frames will essentially make the automobile crash resistant as well as decrease the fuel consumption due to reduced weight. This will also ensure reduction in emissions (pollution) per mileage covered.

- Nanoscale metal oxide ceramic catalysts can aid reduction of polluting particles in the emission gas. Nanocatalysts and membrane technologies will play a critical role in making fuel cells economically viable for replacing the internal combustion engine.
- Nanopowders and coatings can be used to enhance the durability of paint coatings.
- Materials characterization methods, online sensors for the measurement of wear and abrasion, or additives for the adhesion of parts and layers are also being explored. Further developments target the avoidance of lubricants via thin layers on bearings and gliding elements, new electrostatic filters or high power switches in ignition devices via field emission principles, and new catalyst surfaces via highly porous and chemically selective surfaces. Nanoparticles are being used as abrasives, and in paints and electrochromic coatings for windscreens and windows.
- High-efficiency solar cell, hydrogen storage for fuel cell, etc., are being developed as environment friendly fuel resources to drive future automotives.
- High-sensitivity nano-enabled IR sensors for improved night vision.
- Magnetic nanofluids are being developed to improve the efficiency of shock absorbers. It is possible to change the viscosity of the magnetic nanofluid by the application of a magnetic field. Dynamic damping can be achieved by varying the viscosity dynamically. Shock absorbers based on magnetic nanofluids are expected to bring about significant improvement in vibration control compared to traditional shock absorbers.
- Nanoparticle dispersions will enhance the thermal conductivity of fuel.

4.9 WATER TREATMENT AND THE ENVIRONMENT

Water is one of the most essential resources for living things. Although 75% of the earth's surface is covered with water, only 2.5% of the same is potable. Further, less than 1% of freshwater is accessible to humans. With extensive industrialization, the per capita water requirement has increased enormously. To produce one ton of steel about 215,000 litres of water will be needed. Hence, the need for effective water management.

Although several conventional water treatment technologies like chemical treatment, mechanical separation, ultraviolet radiation, biological treatment and desalination are currently available in the market, nanotechnology-enabled products are expected to have a few distinct advantages for large-scale application, to economise and perform with greater efficiency.

Nanotechnology can contribute to effective water management in many ways:

- Nano-membranes and nano-clays for water filtration and desalination
- Nanoparticle-activated wastewater reuse systems
- Nanosensors to monitor water quality against bacteria, heavy metals and toxins

Nano-enabled water treatment technologies include the use of nano-membranes and filters based on carbon nanotubes, nano-porous ceramics, magnetic nanoparticles and other nanomaterials. It has been shown that nano-filtration can lead to remediation of even brackish water. Carbon nanotube–based water filters have been found to be effective tools for nano-filtration. Carbon nanotubes function as molecular filters and allow water molecules to pass through them. Molecules that are bigger than the diameter of the CNT are filtered out. Also, as a consequence of their electronic state, a few smaller ions are also not permitted to enter through the carbon nanotube. Carbon nanotube membranes can reduce the cost of desalination sigificantly. It is reported that the use of nano-titanium dioxide and magnetic nanoparticles can decompose organic pollutants and remove salts and heavy metals, enabling wastewater reuse.

Different molecules can be separated based on their size during nano-filtration using membranes. The technique is mainly applied for the removal of organic substances, such as micropollutants and multivalent ions. In industrial processes, nano-filtration is applied for the removal of specific components, such as colouring agents.

Other applications of nano-filtration are:

- Removal of pesticides from groundwater
- Removal of heavy metals from wastewater
- Wastewater recycling in laundries
- Water softening

There are several conventional technologies in practice today to remove bioorganisms, toxins and impurities from water. It is well known that Saudi Arabia produces nearly 70% of its potable water by the desalination technology. Nanotechnology is expected to result in economic solutions capable of reaching a wider cross section of people in the longer run. For effective remediation of contaminated water, particularly for removing heavy metal ions, various nanoparticles and nanomaterials like zeolites, carbon nanotubes, self-assembled monolayers on mesoporous supports (SAMMS), biopolymers, single-enzyme nanoparticles, zero-valent iron nanoparticles, bimetallic iron nanoparticles and nanoscale semiconductor photocatalysts are in use.

Removal of pathogens from water is essential to avoid several waterborne diseases. It is believed that the use of nanomaterials like silver and titanium dioxide with antimicrobial characteristics can provide a viable alternative to the use of chlorine treatment. Air pollution can be monitored using nanotechnology with filtration. Nano-filters could be applied to automobile and other exhausts in industry to filter out contaminants before the exhaust gases are let into the atmosphere, so that the build-up of greenhouse gases in the atmosphere is prevented. Nanocatalysts can be used in catalytic converters in automobiles to remove contaminants and increase driving performance. Nanosensors could also be developed to detect toxic gases at very low concentrations in the atmosphere.

4.10 NANO-MEDICAL APPLICATIONS

Nanotechnology is promising to revolutionise healthcare technologies in a more patient-friendly direction. Nanotechnology and nanomaterials are being used in diagnosis, therapy and prevention. Significant developments have been illustrated in the field of nano-enabled targetted drug delivery, cancer and TB therapy, disease diagnosis, biosensors for health monitoring, surgical tools, implant materials, tissue engineering, molecular imaging, biodetection of disease markers, etc. Nanotechnology has also enabled the production of 'laboratories on a chip' that perform multiple medical tests (in vitro or in vivo).

When these particles function as nano-medibots that release anti-cancer pharmaceuticals into the cells or penetrate the tissue and deconstruct them mechanically, then they treat cancer. Further, particles may absorb infrared radiation, which is converted to heat to ablate the target (cancer) tissue. Finally, when administered prophylactically (as a nano-vaccination), they can also prevent cancer.

Nano-pharmacology is the use of nanotechnology in pharmacology applications—assembly of current molecular entities; exploring and matching specific compounds to particular patients for maximum effectiveness; and advanced molecular compound delivery systems. Nanoparticles may yield targetted and sustained delivery of pharmaceuticals to specific tissues with a minimum of systemic side effects. In case of nanoshells, manipulating the ratio of wall to core dimensions, they can be precisely tuned to scatter or absorb any particular wavelength of light. Gold-coated nanoshells could convert light into heat, enabling the destruction of tumours.

Nano-pharmaceuticals and nanotechnology drug delivery systems provide greater and more controlled pharmaceutical uptake in tissues throughout the the body. This is critical for oncological applications. 'Trojan Horse' capsules can be used to sneak in a biological compound payload through the blood–brain hurdle to treat ophthalmologic diseases such as macular degeneration, glaucoma and diabetic retinopathy. Nano-pharmacology delivery systems are not limited to internal use—they also enable the absorption of pharmaceuticals through nano-emulsions spread on the skin. Newly developed contrast dyes make it possible to examine patients at a molecular level. Nanotechnology can also be used to partially repair neurological damage. For example, it can improve the correctness of cochlear implants that turn sound into electrical impulses and assemble light-activated implants in the retina to partially restore lost vision. Biomemetic scaffolds are used to support damaged nerves to regrow and reconnect.

4.11 TEXTILES

There are many novel applications of nanotechnology in the textile industry to provide multifunctional attributes to fabrics. For example, the fabric can be made stain resistant, water repellent or absorbing, light emitting, antibacterial, release fragrance in a controlled manner, etc. Antimicrobial properties have been imparted to fabrics by incorporating suitable nanoparticles into nylon and other fabric polymers. It is also possible to coat nanocrystalline zinc oxide particles on synthetic fibres, to impart antimicrobial effect without much change in the colour and gloss of the fabric. Plasma technology is being used to modify the top few (nanometres) layers of textiles, allowing them to be made antibacterial, antifungal and water repellent. Other areas of interest include heat resistance and mechanical resilience to work wear, ballistic protection, sensors and camouflage.

Application of nanotechnology has enabled the development of *intelligent textiles* that are capable of sensing the environment or health of the personnel, to change colour in response to stimuli, and to generate heat. The major motivation for developing intelligent textiles is again perhaps nature. The skin is distributed with a sensor network, to detect pressure, heat, etc. The skin sweats on a hot day to cool the body, and enforces blood circulation on a cold day. Today, scientists have succeeded in developing fabrics with fibres coated with a variety of nanomaterials that can be used as sensors. Fibre sensors, which are capable of measuring temperature, strain/stress, gas and smell, are typically *smart fibres* that can be directly applied to textiles. These are expected to find use in skiwear, shoes, sports helmets and insulation devices. Development of textiles impregnated with sensors that are integrated with global positioning system (GPS) can help the wearer navigate to the desired destination. Fabrics and composites integrated with optical fibre sensors have been used to monitor the soundness of major bridges and buildings. The first generation of wearable motherboards has been developed, which have sensors integrated inside garments and can detect information regarding injury to the health of the wearer, and transmit such information remotely to a hospital.

Shape memory polymers exhibit much higher recoverable strain limits (~100%) in contrast to shape memory metallic alloys. These are expected to have potential application in non-invasive surgery. Intelligent textiles that can change colour and provide camouflage are being developed for military applications. Scientists have developed an innovative process to combine extremely thin layers of two materials: plastic and glass. This results in a new fibre that can reflect all the light that hits it, from any direction. Uniforms woven from these fibres have an optical bar code that will help soldiers distinguish friend from foe on night patrol, or during the smoke and confusion of an attack with firearms.

4.12 PAINTS

Incorporating nanoparticles in paints could improve their performance, for example, by making them lighter and giving them different properties. Thinner paint coatings can reduce the weight. The solvent content of paints may also be substantially reduced with nanopaints. New types of fouling resistant marine paint could be developed and are urgently needed as alternatives to tributyl tin (TBT), now that the ecological impact of TBT has been recognised. Anti-fouling surface treatment is also valuable in process applications such as heat exchange, where it could lead to energy saving. If they can be produced at sufficiently low cost, fouling-resistant coatings could be used in piping for domestic and industrial water systems. It remains speculative whether effective anti-fouling coatings could reduce the use of biocides, including chlorine. Other novel, and more long-term, applications of nanoparticles might lie in paints that change colour in response to change in temperature or chemical environment, or paints that have reduced infrared absorptivity and therefore reduced heat loss. Owing to concerns about the health and environmental impacts of nanoparticles, the durability and abrasive behaviour of nano-engineered paints and coatings have to be addressed, so that abrasion products take the form of coarse or microscopic agglomerates rather than individual nanoparticles.

4.13 ENERGY

Nanomaterials are bound to find a place in green energy technologies too. The most common nanostructured energy enabling technologies that are emerging are:

- Nanostructured photovoltaic systems
- Nanostructured fuel cells
- Hydrogen storage systems
- Efficient light emitting devices

Conventional solar cells suffer from the limitations of poor efficiency and high cost compared to other large-scale energy resources. Nanomaterials-based solar panels can have increased performance and the technology could soon compete with conventional power plants.

The three main types of nano-solar cells being developed are:

- Flexible polymer-based photovoltaics
- Nanoparticle solar cells
- Sprayable self-assembling photocells

Scientists are developing cheap and easy-to-apply plastic solar cells, composed of tiny nanorods dispersed in a polymer that can be easily applied to any surface. These nanorods absorb a particular wavelength to generate electrons. Their efficiency is presently only about 2%, though they are much cheaper. However, it is expected that by tuning the dimensions of the nanotube, it is possible to absorb a wide range of energy from light which would result in improved efficiency. It has been observed that the addition of a small fraction of carbon nanotubes to nanocrystalline TiO_2 film almost doubles the efficiency.

Another promising area for the application of nanotechnology is the 'hydrogen economy'. Hydrogen can be a good alternate fuel of the future not only because it is readily available in water but also because it is a non-polluting source of energy. Low-cost techniques of obtaining hydrogen from water may be worked out with the help of nanocatalysts. Nano-pyramids of (5–15 nm) iridium have been found to be highly efficient in aiding hydrogen generation from ammonia.

Nanotechnology can help in the development of the following:

- Nano-engineered hydrocarbon carbon surface membranes
- Spray depositing platinum on porous alumina
- Replacement of platinum catalysts using less expensive nanomaterials

Nanotechnology has also resulted in a significant about turn in the performance of of Li ion batteries. Application of certain nanomaterials has resulted in a decrease in organic electrolyte reduction during recharging. Nanoparticles quickly absorb and store vast quantities of lithium ions, without causing any deterioration in the electrode. This nano-enabled Li battery is also 60 times faster than the typical lithium ion batteries that are widely used today. In addition, the battery has a long life cycle, losing only 1% of capacity after 1,000 cycles of discharging and recharging, and can operate at very low temperatures. At –40°C the battery can discharge 80% of its capacity, against 100% in an ambient temperature of 25°C. This speedy and highly effective recharge characteristic of the battery will support CO_2 reduction resulting in eco-friendly batteries.

Armchair quantum wire (AQW) is a SWNT-based wire with physical properties conducive to flowing electrons that AQW cables could literally let electricity (in the form of electrons) glide across a grid for 1,000 miles with virtually no resistance or power loss. In comparison to superconductors, the SWNT-based AQWs have an advantage as they do not have to be cooled to cryogenic temperatures. According to Dr Adams,"The armchair quantum wire can simply by its structure propagate an electron down the length of a nanotube, much like light waves travel down an optical fiber." The AQW cables would be a revolutionary leap beyond copper, providing lighter, stronger and more conductive cables with vastly more capacity. AQWs are tiny, and about 1014 wires would be present in 1 cm. Although each AQW can conduct only 20 μA of electricity, it would be possible to bundle 10–14 of them together to get 10 million amps down a single filament. It should also be possible to have multiple filaments together in a single cable, like multiple filaments in an optical cable, resulting in a capacity of over billions of amps, in contrast to copper cables that carry about 2000 amps.

Many significant efforts are in progress to identify and utilise new energy sources, to increase the production of existing sources, to increase conversion and storage efficiency, and, equally important, to reduce pollution. In MIT's Laboratory for Electromagnetic and Electronic Systems, they are exploring a nanostructured ultracapacitor electrode that has the potential to increase a capacitor's energy storage density to approach that of a chemical battery. Nanostructured emissive coatings and filters that significantly increase the efficiency of direct thermophotovoltaic (TPV) generation of electricity from heat are also under development.

4.14 DEFENCE AND SPACE APPLICATIONS

Nanotechnology can play a vital role in defence and space systems by minimising size, weight and power consumption that are important for long-range coverage. Also, embedded nanosensors enabled with wireless networks can enhance the intelligence of future warfare systems. There are several possible applications of nanotechnology for defence and space applications, as shown below:

- Light-weight vehicles to enhance fuel economy; longer distances can be covered for the same fuel capacity
- Smart components with built-in condition and load-monitoring sensors, such as fibre Bragg
- Adaptive structures, like adaptive skin, for better thermal control
- High energetic propellants such as nano-dispersed aluminum
- Lightweight protective clothes
- Anti-ballistic and shatterproof armour
- Advanced sensors, such as, high-resolution vision systems, RF, infrared, acoustic arrays, terahertz and through-the-wall radar vision
- Physical identification tags (RFID) for goods, digital ID tags for documents and information
- Biometric sensing for personal identification characteristics, such as, fingerprint, face, DNA

- Ambient intelligence for surveillance by means of distributed wireless sensor networks
- Tracking and tracing with position and motion sensors

Intelligent textiles with embedded sensors in the soldiers' attire is another important potential area for application of nanotechnology. The clothes soldiers wear can be modified to provide better protection against heat and cold and integrated nanosensors can be used for remote monitoring of heartbeat and blood pressure. In addition, clothing can have built-in protection against bullets.

Nanotechnology is needed for improved detector sensitivities (signal-to-noise ratios), to miniaturise sensor arrays for selectivity, to tailor-make high surface area materials for detection/absorption/deactivation and to create selective catalysts. Microsystems technology and nanotechnology will therefore enable small portable sensor systems capable of identifying chemical, bio, nuclear, radiation or energy threats. This will enhance the flexibility of deployment and operations, increase the safety of soldiers and civilians and enhance environmental security. Onboard intelligence will continuously increase, facilitating automated control and maintenance. Naval vessels have additional requirements with respect to detection and surveillance sensor systems.

Other technologies which can be integrated in future weapons are RF ID-tags in guns, cartridges, target positioning/recognition (via micro IR camera on guns or PDAs, microradar, RF-array, through-the wall THz radar), and smart helmets to pull the trigger from a distance (teleweapon), various ammotypes (shaped ceramic materials, softer bullets, high-penetration bullets, sensor modes/smart dust and insensitive tailored explosives to limit collateral damage).

Closely connected to the concept of the smart helmet is that of *anti-ballistic protection* via a light-weight helmet consisting of a combination of polymers, fibres and nanomaterials. This new nanocomposite should have higher impact resistance than present fibre composite systems and should have a significantly reduced weight. The application of nanofibres and buckypaper in combination with present high-strength fibres and polymer materials to create new composite materials seems to be most promising. For the next generation of helmets, the use of fibres such as Kevlar, Dyneema or M5 is most realistic in combination with nanomaterials which can enlarge the strength of the composite and keep the fibres closely packed in the composite structure at impact. M5 is the newest type of fibre in this category and product applications of this fibre are expected in the next few years. Nanoclay in a silliputty-type of polymer matrix combined with fibres is a possible alternative. These high-strength fibres can, in theory, bring significant improvement. The same applies to metal nanoparticles coated with a multiple-layered ceramic nanocoating. These coatings were originally developed for turbine blade protection but the extreme hardness is also of use for anti-ballistic nanofillers in composites. In the long term (10–15 years), a more dominant use of electrospun nanofibres to create basic fibre strength in the composite can be expected. These CNT fibres and other nanofibres have a theoretical strength of 130–180 GPa. Thanks to the technological advances in high strength polymer fibres such as carbon, aramide and Dyneema, the performance of antiballistic suits has improved considerably over the last 20 years, with subsequent reduction in weight (30%). These suits with integrated or inserted composite fibre structures are quite effective and are being successfully applied for ballistic protection of the body. However, the composite structures are not sufficiently flexible to be used for protection of extremities such as the arms, legs and neck. At present, injuries of these extremities have become the dominant factor in casualties, especially from bombings and subsequent shatter, resulting in loss of military power and high cost of medical treatment. In view of this, several concepts for flexible armour have been proposed and are now in development.

Magneto-restrictive fluid—a nanoparticle-filled flexible medium that can be electrically activated to become rigid

Shear-thickening fluid—a nanoparticle-filled binder for high-strength textiles that is flexible under low shear rate and that becomes rigid under high shear rate impact (passive system, ARL). This nanoparticle-filled system inhibits deformation and sliding of high strength fibres at high shear rate.

Silliputty-type of elastomers in combination with ceramic armour—an elastomer system which is deformable and elastic at low shear rate and stiff at high shear rate. It is similar to shear thickening fluids, but has so far been less effective in antiballistics (passive system, for example, D_3O material).

The ultimate goal will be to create nanorobots or nanobots for activities on land and NUAV (nano-unmanned aerial vehicles) for reconnaissance and sensory activities in the air (flying artificial insects). Uninhabited combat vehicles (fighter, submarine, and vehicle) with higher performance and lower casualty risk can also be expected. Ideally, all microvehicles and robots should be less visible for enemy troops (biomimetic structures), should last long enough to gather essential information, and should be low in cost and therefore redundant.

Wounds can be dressed with *intelligent bandaids* which monitor the moisture level and bacterial activity, and release antimicrobials on nanoparticles to kill the bacteria. Part of this health monitoring system can be a portable sample preparation and lab-on-chip analysis kit enabling the soldier to test his own body fluids when he needs more specific data and water and food. Ideally, this analysis system will be built into his smart suit and will be able to detect bioagents and apply antidotes to the soldier. Biosensors for body-function monitoring can be expected to be integrated subcutaneously in the body of future combat soldiers. Core temperature measurement with a swallowable passive RF-sensor is one of the solutions for accurate and low-cost core temperature measurement of soldiers in combat.

Smaller satellites are now becoming feasible, owing to the miniaturization enabled by nanotechnology. Besides reducing their size, weight and power consumption, the use of micromachined devices could give better component integration in areas such as propulsion, communication, data processing, power generation and navigation. With a distributed network of small satellites instead of one big one, both functionality (multi-aperture synthesis for better accuracy, formation flying) as well as redundancy is gained. The ultimate goal is to develop a complete satellite-on-chip, the so-called picosatellites. Microsatellites (~ 10–30 cm) are in development in many countries: USA/NASA, UK, France, Germany, Sweden and Spain, both for civil and defence applications.

4.15 STRUCTURAL APPLICATIONS

The increased rate of forming or the lower temperature related to the superplastic deformation of nanocrystalline materials would make superplasticity more industrially accessible, extending its possible limits of use. With nanocrystalline metals, superplastic deformation can extend rapid and large-scale forming processes. It is speculated that nanocrystalline superplasticity will have an advantage over traditional superplastic materials when materials chemistry may not be changed because of the nature of the application (for example, in electronics applications), or high strength is demanded after forming. Another area is that of diffusion bonding. It has been shown that the use of a superplastic intermediate layer in *diffusion bonding* of non-superplastic stainless steel dramatically improves the properties of the joint, especially if the mating surfaces are rough. Superplasticity may also be utilised in the processing of nanocrystalline ceramics themselves. Nanocrystalline ceramics are difficult to produce by the pressureless sintering routes typically used for conventional ceramics, but they may be produced by sinter forging which uses the superplasticity of the materials itself by closing pores with the aid of plastic flow.

The higher strength of nanocrystalline materials may be utilised in several potential applications, when processing of the materials is adequately developed. The development of nanocrystalline M50 steel as the main shaft bearing material has improved the performance of engines in gas turbine industries. Development of WC–Co nanocomposites has been driven by the expectation of obtaining cutting tools and hard metal coatings with superior properties compared to their traditional counterparts. These materials are already starting to have commercial impact and are used in the manufacture of machine tools, drill bits and wear parts. Tools made of cemented carbide nanocomposites have enhanced hardness, fracture toughness and wear resistance compared to their conventional counterparts.

Currently, *nanocrystalline titanium* is considered to be a potential material for medical implants. In order to obtain adequate strength, titanium alloys (mainly Ti–6Al–4V) are used, for example, for hip prostheses. Development of nanocrystalline pure titanium for such an application would allow the alloying content to be decreased for increased biocompatibility. Both the static and the fatigue strength of commercially pure titanium fasteners and threaded articles could be substantially increased by ECA processing, producing a nanocrystalline grain structure. Nanocrystalline ceramics have also been considered for orthopedic and dental implants of the future. Nanomaterials with improved mechanical properties could then replace some of the conventional biomaterials and could be tailored to meet clinical requirements associated with anatomical differences or patient age.

Incorporation of nanotubes instead of carbon fibres as reinforcing elements into plastic, ceramic and metallic matrixes can potentially provide structural materials with dramatically improved modulus and strength. Many improvements have taken place in the use of nanoparticles as filler materials in polymers. These include fillers in dental polymers to improve their performance, for example, wear resistance or stiffness as in polymer-layered silicate nanocomposites.

Applications of nanocrystalline metallic materials have been limited because of low ductility in tension. However, cold rolling of nanocrystalline copper has opened up interesting vistas for developing novel processing of some metallic materials utilizing the nanocrystalline structure. The traditional deformation–annealing technique routinely used may be much simplified by using nanocrystalline metals as starting materials. With proper post–heat treatments, microstructure may be easily controlled so that the desired properties in the final product can be achieved.

SUMMARY

Nanomaterials and technology are making their impact in almost all areas of life.

- Development of novel devices based on nano-opto-electronic materials, molecular devices and quantum structures are creating new directions for miniaturization with increased efficiency of nano-electronic systems.
- MEMS and NEMS find extensive use both as sensors and actuators in a wide spectrum of engineering application.
- An entire gamut of nanosensors for biomonitoring, health parameter surveillance, safety logics, environmental control, process control, etc., have been developed.
- In the field of medicine, nanotechnology finds application for both diagnostic tools as well as for advanced therapy.
- Development of smart textiles with in-built sensors and functional nanoparticles are set to be introduced on a wide scale both for defense as well as for domestic use.
- Nanomaterials have a promising future in enhancing efficiencies of green energy technologies, like solar cells, hydrogen cells, etc.

EXERCISE

1. Define: (a) Moore's law, (b) Spintronics and (c) Quantum effect devices.
2. Write a note on the application of nanotechnology for energy production, storage and enhancing energy efficiency of buildings/appliances.
3. What are molecular devices? What are their advantages over conventional microelectronic devices?
4. Explain the possible applications of CNTs for pressure and gas sensor applications, with mechanisms of their working.
5. Discuss the role of NEMS technology in advanced engineering applications.
6. Write a note on the use of nanometallic particles as catalysts.
7. Discuss the role of nanotechnology in water treatment and environment control.
8. Explain the mechanisms and fabrication of at least three different biological sensors.

9. Discuss the various possible applications of nanostructured silver, titanium oxide and zinc oxide.
10. Explain at least two applications of hard nanostructured coatings.

Gordon Earle Moore is the co-founder of Intel Corporation, USA. Born in 1929 at San Francisco, California, he obtained his doctoral degree in Physics and Chemistry from CalTech in 1954. Intel Corporation was co-founded by Moore in 1968, where he served as Executive Vice President until 1975, after which he became President and Chief Executive Officer. Dr Moore became Chairman of the Board and Chief Executive Officer in 1979. He became Chairman Emeritus of Intel Corporation in 1997.

Chapter 5

Tools to Characterize Nanomaterials

Learning objectives

- Structural, microstructural and microchemical analysis of nanomaterials using X-ray diffraction and electron microscopy
- Surface topography profiling of nanostructures using atomic force microscopy
- Characterization at atomic resolution using 3D atom probe and field ion microscope techniques
- Nano-mechanical surface property measurement using nanoindentation

The characterization of small structures or small-sized materials in the nanometric-scale usually calls for sophisticated characterization tools. Characterization of nanomaterials and nanostructures has been largely based on certain critical advancement of conventional characterization methods developed for bulk materials. For example, X-ray diffraction (XRD) has been widely used for the determination of crystalline character, crystallite size, crystal structures and lattice constants of nanoparticles, nanowires and thin films. Scanning electron microscopy (SEM) and transmission electron microscopy (TEM), together with electron diffraction, have been commonly used in the characterization of nanoparticles to get an idea of the size, shape and defects present in these materials.

Optical spectroscopy is used to determine the size of semiconductor quantum dots. Scanning probe microscopy (SPM) is a relatively new characterization technique and has found widespread application in nanotechnology. The two branches of SPM are scanning tunnelling microscopy (STM) and atomic force microscopy (AFM). Although both STM and AFM are true surface imaging techniques that can produce topographic images of a surface with atomic resolution in all three dimensions, in combination with appropriately designed attachments, STM and AFM have found a much broader range of applications, such as nanoindentation, nano-lithography and patterned self-assembly. Almost all solid surfaces, whether hard or soft, electrically conductive or not, can be studied with STM/AFM. Surfaces can be studied in a gaseous medium such as air or vacuum, or in liquid. In this chapter, we will briefly discuss these characterization techniques and their applications in nanotechnology.

Characterization and manipulation of individual nanostructures require not only extreme sensitivity and accuracy, but also atomic-level resolution. It therefore leads to various microscopy techniques that will play a central role in the characterization and measurement of nanostructured materials and nanostructures. Miniaturization of instruments is obviously not the only challenge. The new phenomena—physical properties and short-range forces—which do not play a noticeable role in macroscopic-level characterization, may have significant impact at the nanometric-scale.

5.1 X-RAY DIFFRACTION (XRD)

XRD is extensively used to study the crystal structure of solids, defects and stresses. In XRD, a beam of X-rays, with wavelength ranging from 0.07 to 0.2 nm, is diffracted by the crystalline specimen according to Bragg's law:

$$\lambda = 2d \sin \theta$$

where d is the interplanar distance and λ is the wavelength of the X-rays. The intensity of the diffracted beam is measured as a function of the diffraction angle (2θ) and the specimen's orientation. The diffraction pattern can be used to identify the crystalline phases and their structural characteristics. XRD is non-destructive and does not require detailed sample preparation.

Homogeneous and inhomogeneous strains in materials can be measured as X-ray intensity and is dependent on the Bragg angle. Homogeneous or uniform elastic strain shifts the diffraction peak positions, without change in peak profile. A shift in the X-ray peak positions indicates a change in d-spacing caused by a change in lattice constants. Inhomogeneous strains vary from crystallite to crystallite or even within a single crystallite. Since XRD gives averaged information from all such crystallite volumes, it leads to broadening of the diffraction peaks, which increases with increase in $\sin \theta$. Peak broadening can also be due to the fine crystallite size, which is independent of $\sin \theta$. The contribution of crystallite size and lattice strain to peak broadening can be independently determined by peak profile analysis.

In the absence of inhomogeneous strains, the crystallite size, D, can be estimated from the peak broadening using Scherrer's formula:

$$D = \frac{K\lambda}{B \text{ Cos } \theta_B}$$

where λ is the X-ray wavelength, B is the full width at half maximum (FWHM) height of a diffraction peak, θ_B is the diffraction angle, and K is Scherrer's constant, which is of the order of unity for a spherical crystal. However, nanoparticles often form twinned structures, and hence Scherrer's formula may not always give true particle sizes. It is also important to note that X-ray diffraction provides only an average crystallite size. The thickness of epitaxial and highly textured thin films can also be determined using XRD.

Due to lower energy of the X-ray beam used, the X-ray diffraction intensities are low, particularly in case of low atomic number materials and hence, detection of phases with small volume fractions is difficult with XRD. Electron diffraction intensities are usually $\sim 10^8$ times larger than that for XRD. Figure 5.1 shows the XRD patterns of unmilled and 20-hour milled Cu powder and that of mechanically alloyed NiCoCrFe equiatomic blend showing significant peak broadening after milling, indicating the formation of nanocrystallites.

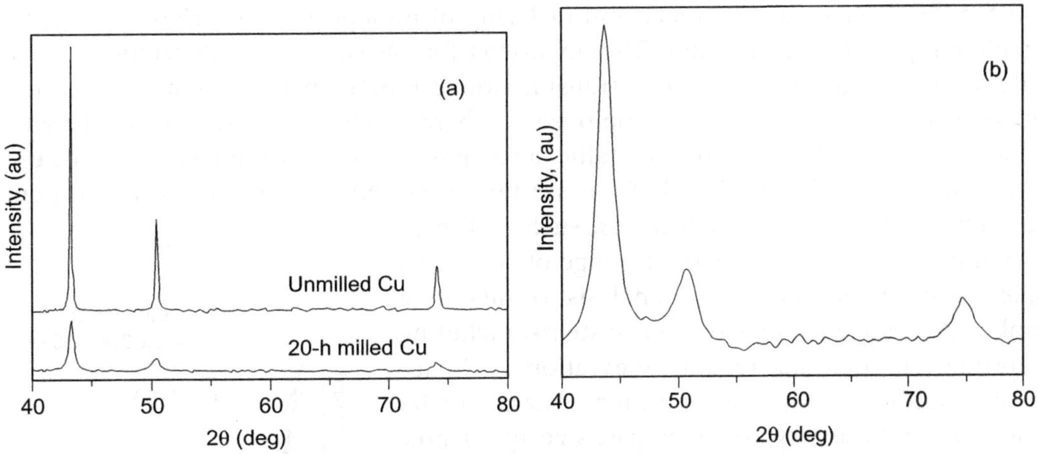

Fig. 5.1 XRD patterns of nanocrystalline (a) copper and (b) NiCoCrFe equiatomic alloy obtained by high-energy ball milling (*Source:* BS Murty, IIT Madras).

5.2 SMALL ANGLE X-RAY SCATTERING (SAXS)

SAXS is another powerful tool for characterizing nanostructured materials. Strong diffraction peaks result from constructive interference of X-rays scattered from ordered arrays of atoms and molecules. A variety of information can be obtained from the angular distribution of scattered intensity at low angles. Fluctuations in electron density over lengths in the order of 10 nm or larger are sufficient to produce appreciable scattered X-ray intensities at angles of $2\theta < 5°$. These variations can be due to differences in density or composition, or both, and do not need to be periodic. The amount and angular distribution of scattered intensity provides information about the size of very small particles or their surface area per unit volume, regardless of whether the sample or particles are crystalline or amorphous.

For example, consider a composite structure of two phases separated by well-defined boundaries, such as nanoparticles dispersed in a homogeneous medium. The variation in electron density can arise due to variation in atomic structure or local variation in their chemistry. SAXS is the scattering due to the existence of regions of inhomogeneous sizes ranging from several nanometres to several tens of nanometres, whereas XRD is used to determine atomic structures of the crystallite phases. In essence, SAXS probes relatively large-scale structures, in contrast to wide angle X-ray diffraction (WAXD) that deals mainly with the atomic structure of crystals. SAXS includes not only the diffraction from large lattice spacing of the order of tens, hundreds or even thousands of interatomic distances, but also the scattering by perturbed or non-periodic structures of amorphous and mesomorphic materials.

Since SAXS is very effective in measuring inhomogeneity in the region of 1–100 nm, it has been widely used in the characterization of nanocrystals. SAXS has also been widely used for the determination of size and ordering of mesoporous materials synthesized by organic-templated condensation. The apparatus for measuring the distribution of small angle scattering generally employs transmission geometry using a fine monochromatic radiation beam. SAXS has been employed to characterize a wide spectrum of materials including biological structures, metallic and non-metallic specimens, composites and mesoporous materials. Small angle X-ray scattering (SAXS) is a useful analytic technique for the study of aggregated nanopowders, such as fumed silica and titania. If a wide enough range of scattering angles is observed, the SAXS analysis results in a number of average features of such systems, including the primary particle size (radius of gyration), primary particle surface area, concentration, mass fractal dimension of the aggregates, aggregate size and degree of aggregation.

The size distribution of sub-nanoclusters, which cannot be easily estimated using TEM, can successfully be analysed using SAXS. In one study, the size of icosahedral particles formed by the crystallization of a Zr–Pd metallic glass was studied using SAXS. The SAXS profiles in Fig. 5.2 have been analysed and the results show that the size of icosahedral spherical particles is 6.5 nm and 6.7 nm, respectively, at 715 K and 725 K, which is very close to that measured by TEM. Small angle X-ray scattering has also been used for analysing the structure of polymers (electrospun polymer nanofibres), especially with the use of synchrotron radiation.

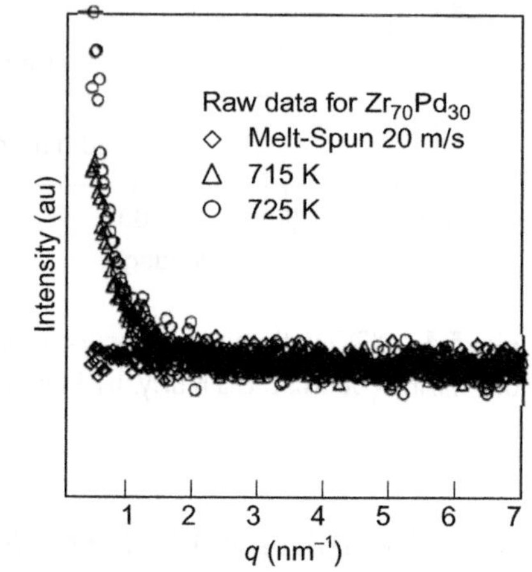

Fig. 5.2 SAXS profile of $Zr_{70}Pd_{30}$ in the as melt-spun amorphous state and that after crystallization to nanocrystalline icosahedral phase at two different temperatures. (*Source:* BS Murty, IIT Madras).

5.3 SCANNING ELECTRON MICROSCOPY (SEM)

Scanning electron microscopy (SEM) is one of the most popular and widely used techniques for the characterization of nanomaterials and nanostructures. SEM can be effectively used to characterize specimens down to a resolution of a few nanometres, with image magnifications achievable in the range of ~10 to over 300,000. In addition to information on surface topography, SEM can also provide useful information on chemistry, crystal orientation and internal stress distribution. SEM consists of an electron gun to emit electrons that are focussed into a beam, with a very fine spot size of ~5 nm. Electrons are accelerated to energy values in the range of a few hundred eV to 50 KeV, and rastered over the surface of the specimen by deflection coils. As the electrons strike and penetrate the surface, a number of interactions that result in the emission of electrons and photons from the sample occur, and SEM images are produced by collecting the emitted electrons on a cathode ray tube (CRT). Various SEM techniques are differentiated on the basis of what is subsequently detected and imaged. The principle images produced in SEM are of three types: secondary electron images, backscattered electron images and elemental X-ray maps.

5.3.1 Electron–matter interaction

When a high-energy primary electron interacts with an atom, it undergoes either inelastic scattering with atomic electrons or elastic scattering with the atomic nucleus. In an inelastic collision with an electron, the primary electron transfers part of its energy to the other electron. When the energy transferred is large enough, the other electron will emit from the sample. If the emitted electron has energy of less than 50 eV, it is referred to as a *secondary electron (electron emitted from one of the orbitals of the incident atom)*. Since the secondary electron energy is small, the SE images are highly sensitive to topographic variations. *Backscattered electrons* are the high-energy incident electrons that are elastically scattered and essentially possess nearly the same energy as the incident or primary electrons. The probability of backscattering increases with the atomic number of the sample material. Although backscattering images cannot be used for elemental identification, useful contrasts can develop between regions of the specimen that differ widely in atomic number, Z. Hence the BSE image can provide atomic number contrast in addition to topographic contrast.

An additional electron interaction in SEM is that the *primary electron* collides with and ejects a *core electron* from an atom in the sample. The excited atom will decay to its ground state by emitting either a characteristic X-ray photon or an Auger electron, both of which have been used for chemical characterization. The energy of the emitted characteristic X-ray or Auger electrons are specific to the chemistry of the incident atom and hence study of these can provide useful information on the chemistry of the sample investigated. By focussing the electron beam to fine points, it is possible to get localized information on chemistry variations. However, it should be remembered that such information on characteristic X-ray (and hence information on chemistry) is generated from a depth of approximately a micron and hence the information is averaged over this depth although the beam size could be finer on a lateral scale. Combined with chemical analytical capabilities, SEM not only provides an image of the morphology and microstructures of bulk and nanostructured materials and devices, but can also provide detailed information of chemical composition and distribution.

5.3.2 Imaging

The theoretical limit to an instrument's resolving power is determined by the wavelengths of the electron beam used and the numerical aperture about the system. The resolving power, *R*, of an instrument is defined as:

$$R = \frac{\lambda}{2NA}$$

where λ is the wavelength of electrons used and *NA* is the numerical aperture, which is engraved on each objective and condenser lens system, and a measure of the electron gathering ability

Fig. 5.3 HRSEM image of nanocrystalline (a) $ZnTiO_3$ and (b) $Ni_{0.8}Zn_{0.2}Fe_{1.98}O_{4-\delta}$. (*Source:* BS Murty, IIT Madras).

of the objective, or the electron providing ability of the condenser. Figure 5.3 shows the SEM images of zinc titanate and NiZn ferrite nanoparticles prepared by high-energy ball milling.

The different aspects of crystallite information that can be investigated using SEM are listed below:

Topography: The surface features of an object or 'how it looks'; detectable features are limited to a few manometres.

Morphology: The shape, size and arrangement of the particles making up the object that are lying on the surface of the sample or have been exposed by grinding or chemical etching; detectable features limited to a few manometres.

Composition: The elements and compounds the sample is composed of and their relative ratios, in areas ~ 1 micrometre in diameter and depth.

Crystallographic information: The arrangement of atoms in the specimen and their degree of order; only useful on single-crystal particles >20 micrometres.

A detailed explanation of how a typical SEM functions is given in Fig. 5.4 (see Plate 5).

1. The electron gun produces a stream of monochromatic electrons.
2. The stream of electrons is condensed by the first condenser lens. The condenser lens not only forms the beam but also limits the amount of current in the beam. The condenser lens together with the condenser aperture eliminates the high-angle electrons from the beam.
3. The condenser aperture constricts the beam and eliminates some high-angle electrons.
4. The second condenser lens focusses the electrons into a thin, tight, coherent beam and is usually controlled by the fine probe current knob.
5. An objective aperture further eliminates some high-angle electrons from the beam.
6. A set of coils scan the beam in a grid fashion, the period of time for which the beam dwells on each point is determined by the scan speed (usually in the microsecond range).
7. The final lens, the objective lens, focusses the beam on the desired part of the specimen.

8. When the beam strikes the sample, it interacts with the sample and generates radiations, which are detected with various instruments.
9. Before the beam moves to its next dwell point, these instruments count the number of interactions and display a pixel on a CRT, the intensity of which is determined by this number (the more the reactions, the brighter the pixel).
10. The speed of scanning is such that the entire pattern can be scanned 30 times in a second.

5.4 TRANSMISSION ELECTRON MICROSCOPY (TEM)

Transmission electron microscopy can provide microstructural, crystal structure as well as micro-chemical information with high spatial resolution from each of the microscopic phases individually. TEM is therefore a very powerful tool for materials characterization. TEM studies not only aid in improved insight into structure–property correlations, but also in alloy development for improved performance.

In 1924, de Broglie brought out the wave-character of electron rays, which has been the basis for the construction of the electron microscope. The origin of TEM can be traced to the development of magenetic solenoid coils to focus cathode rays in high-voltage CRT, mainly to understand the effect of high-voltage surges—such as those caused by lightening—on damage to transmission lines. Although Dennis Gabour from the Technical University, Berlin, was the first to use iron shrouded coils to focus cathode rays in the innovative CRT built in 1924–26 towards his PhD, he did not realise their magnetic lens effect until Hans Busch published a paper in 1926 on the effect of short solenoid coils in focussing electrons. Thus, though Dennis Gabour was the first to demonstrate the focussing effect of electrons, Busch is still considered as the father of electron optics.

To observe the image of electrons emitted from a cold cathode on a fluorescent screen, Ruska and Knoll succeeded in magnifying the image by 17X by using a second solenoid coil between the first lens and the final plate. This can be considered as the first prototype TEM to demonstrate the focussing effect of transmitted electrons to form magnified images. In 1935, Knoll, while studying television camera tubes, succeeded in forming a topographic image of solid substances using a scanned electron beam with a resolution of about 0.1 mm, thus establishing the framework for developing the scanning electron microscope.

5.4.1 Advanced techniques of TEM

From the first prototype TEM with an image magnification of 17X, developments in electronics, electron optics and lens fabrication have resulted in tremendous improvements, leading to the current dedicated high-resolution TEM capable of sub-Angstrom spatial resolution with the possibility of recording images of over a million times in magnification. With field emission gun electron sources and by using aberration corrected lenses, resolutions less than 0.1 nm have been achieved. In addition, computational abilities have improved to a point where materials properties can be predicted from computer models that contain a similar number of atoms, as is observable by high resolution TEM (HRTEM).

Transmission electron microscopes are improving rapidly to meet the current requirements of developing advanced materials for strategic applications and also by an equally fast evolution of nanotechnology and semiconductor technology. Apart from lattice imaging by HRTEM to study interfaces, defects and precipitation studies, crystal structure of the individual phases can also be obtained by selected area and microdiffraction techniques in TEM. Convergent beam electron diffraction (CBED) can be used to obtain information on symmetry, lattice, strains, thickness, etc., from localized regions of the crystal. Although symmetry information can also be obtained from X-ray diffraction, the advantage of CBED is that symmetry information of individual phases from localized sub-microscopic regions can be obtained along with their morphology, interface coherency/defect structure and micro-chemistry from associated techniques in TEM. Micro-chemical information can be obtained by either energy dispersive analysis of X-rays (EDAX) or by using electron energy loss spectroscopy (EELS). EELS is a highly sensitive micro-analytical tool with high spatial resolution that can identify low atomic number elements as well. Apart from elemental concentrations, information on electron density, density of states and electronic structure as well as site symmetry, radial distribution functions and specimen thickness can be obtained with high spatial resolution by EELS.

Using a technique called ALCHEMI (Atom Location by CHanelling Enhanced MIcroanalysis) based on electron channelling effect and used in association with analytical techniques like EDAX and EELS, it is also possible to identify individual element sites of occupation in ordered crystals. In fact, current technology has advanced to such a level that it is also possible to remotely operate the microscope from a control room situated away from the actual microscope via a computer supported with a web browser and necessary interface. With the feasibility to obtain information on the entire gamut of material including microstructure, crystal structure, symmetry, lattice strain, interface lattice structure, micro-chemistry, bonding state of constituent elements in the phase, etc., TEM is an inevitable tool in the development and characterization of advanced materials. TEM is commonly used for studying defects in crystals.

In contrast to X-rays, electron beams can be focussed using magnetic fields and can hence be used for imaging. Magnetic fields can be used as convex lenses for the electron waves. As with conventional optical lenses, the lens aberrations are difficult to correct even in magnetic lenses. However, the intensity distribution of the electron waves leaving the specimen can be magnified by an electron optical system and resolutions of < 0.1 nm are possible.

Inelastic and elastic scattering are two ways in which electrons interact with the material. Inelastic scattering (leading eventually to absorption) must be minimized to obtain good image contrast, since it contains no local information. The electron beam then gets elastically scattered (diffraction). The lattice defects modulate the amplitude and phase of the incident and diffracted beams.

In TEM, an electron beam with energy of 100 keV or more (up to 3 MeV) is allowed to go through a thin specimen (less than 200 nm). TEM offers special resolution down to an Angstrom, high magnifications up to 10^6 and it can work as a microscope and a diffractometer. Keeping inelastic scattering of the electrons small has supremacy; this demands specimen thickness between 10 nm and 1 μm. The resolution depends on the thickness; high-resolution TEM (HRTEM) demands specimen thickness in the nanometre region.

5.4.2 Quantitative analysis using TEM

Practically all detailed information about extended defects comes from TEM investigations which not only show the defects but, using proper theory, provides quantitative information about say, strain fields. The electron-optical system not only serves to magnify the intensity (and, in HRTEM, the phase) distribution of the electron waves leaving the specimen, but, at the press of a switch, provides electron diffraction patterns. Figure 5.4 (see Plate 5) shows the basic electro-optical design of a TEM.

By tilting the specimen relative to the incident electron beam, it is possible to satisfy the Bragg condition either for many or just two or none of the reciprocal lattice vectors **g**. The preferred condition for regular imaging is the 'two-beam' case with only one 'reflex' excited—the Bragg condition is only satisfied for one point in the reciprocal lattice or one diffraction vector **g** [usually with small Miller indices, e.g., {111} or {220}]. The main challenge in any TEM investigation is specimen preparation. Obtaining specimens thin enough and containing the defects to be investigated in the right geometry (e.g., in cross section) is a big task. TEMs are patterned after transmission light microscopes and will yield the following information.

Morphology: The size, shape and arrangement of the particles which make up the specimen as well as their relationship to each other on the scale of atomic diameters.

Crystallographic information: The arrangement of atoms in the specimen and their degree of order, detection of atomic-scale defects in areas a few nanometres in diameter.

Compositional information: The elements and compounds the sample is composed of and their relative ratios, in areas a few nanometres in diameter.

5.4.3 Functioning of a TEM

The functioning of a TEM is similar to a slide projector in some respects. A projector shines (transmits) a beam of light through the slide; as the light passes through it is affected by the structures and objects on the slide. This causes only some part of the light beam to be transmitted through some part of the slide. This transmitted beam is then projected on the viewing screen, forming an enlarged image of the slide. A more technical explanation of the working of a typical TEM is as follows (Fig. 5.4, see Plate 5):

1. The electron gun at the top of the microscope produces a stream of monochromatic electrons.
2. This stream is focussed to a small, thin, coherent beam by the use of condenser lenses 1 and 2. The first lens determines the *spot size,* the size of the electron beam that strikes the sample. The second lens changes the size of the spot from a wide dispersed spot to a pinpoint beam on the sample.
3. The beam is restricted by the condenser aperture (usually user selectable), knocking out high-angle electrons (those far from the optic axis, the dotted line down the centre).
4. A part of the beam that strikes the specimen is transmitted.
5. This transmitted portion is focussed by the objective lens into an image.

6. Optional objective and selected area metal apertures can restrict the beam—the objective aperture enhancing contrast by blocking out high-angle diffracted electrons, and the selected area aperture enabling the user to examine the periodic diffraction of electrons by ordered arrangements of atoms in the sample.
7. The image is passed down the column through the intermediate and projector lenses, being enlarged all the way.
8. When the beams strike the phosphor image screen light is generated, which causes the image to be visible to the operator. The area through which very few electrons are transmitted look darker (they are thicker or denser). The lighter areas of the image represent those areas of the sample through which more electrons were transmitted (they are thinner or less dense).

When the electron beam passes through the specimen, its interaction with the electrons present in the specimen generates a variety of radiation, which can give a lot of information about the specimen. Elastic scattering gives diffraction patterns and no energy is lost in this process. Inelastic interaction of the electron beam with the electrons at dislocations, grain boundaries, and second-phase particles, in the sample can lead to changes in the intensity of the transmitted electrons. Both elastic and inelastic scattering can give immense information about the sample.

The high magnification or resolution of TEM is a result of the small effective electron wavelengths, λ, which is given by the de Broglie relationship:

$$\lambda = \frac{h}{\sqrt{2mqV}}$$

where m and q are the electron mass and charge, h is Planck's constant, and V is the potential difference through which electrons are accelerated. For example, electrons of 100 keV energy have wavelengths of 0.37 nm and are capable of effectively transmitting through ~ 0.6 μm of silicon. The higher the operating voltage of a TEM instrument, the greater its lateral spatial resolution. The theoretical instrumental point-to-point resolution is proportional to $\lambda^{3/4}$. High-voltage TEM instruments (for example, 400 kV) have point-to-point resolutions better than 0.2 nm. High-voltage TEM instruments have the additional advantage of greater electron penetration, because high-energy electrons interact less strongly with matter than lower energy electrons. So it is possible to work with thicker samples on a high-voltage TEM. Unfortunately, the depth resolution of TEM is poor. Electron scattering information in a TEM image originates from a three-dimensional sample, but is projected on a two-dimensional detector.

5.4.4 TEM specimen preparation

Preparation of samples that are thin enough for electron transparency is a difficult task; however, it is much easier for nanomaterials. Selected area diffraction (SAD) offers a unique capability to determine the crystal structure of individual nanomaterials, such as nanocrystals and nanorods, and the crystal structures of different parts of a sample. In SAD, the condenser lens is defocussed to produce parallel illumination at the specimen and a selected area aperture is used to limit the diffracting volume. SAD patterns are often used to determine the Bravais lattices and lattice parameters of crystalline materials by the same procedure used in XRD. Although TEM has no inherent ability to distinguish atomic species, electron scattering is exceedingly sensitive to the target element and various spectroscopy techniques are developed for the analysis of chemical composition.

TEM has also been used to determine the melting points of nanocrystals. Heating of a nanocrystal by the electron beam can cause melting, which can be identified by the disappearance of crystalline diffraction. TEM has also been used to measure the mechanical and electrical properties of individual nanowires and nanotubes. Thus, one can develop structure–property correlations in nanomaterials. Figure 5.5 shows nano-quasicrystalline particles in a Zr alloy.

Fig. 5.5 Nano-quasicrystalline particles in Zr-Cu-Al-O alloy. (*Source:* BS Murty, IIT Madras).

High-resolution TEM (HRTEM) is the ultimate tool for imaging atomic-scale defects. In favourable cases, it shows a two-dimensional projection of the crystal with defects and other features. Of course, this only makes sense if the two-dimensional projection is down some low-index direction, when atoms are exactly on top of each other. Consider a very thin slice of a crystal that has been tilted such that a low-index direction is exactly perpendicular to the electron beam. All lattice planes nearly parallel to the electron beam will be close enough to the Bragg position and will diffract the primary beam. The diffraction pattern can be considered as the Fourier transform of the periodic electron potential in crystals. In the objective lens, all diffracted beams and the primary beam are brought together again; their interference provides a back-transformation and leads to an enlarged image of the periodic potential. This image is magnified by the following electron-optical system and is finally seen on the screen at typical magnifications of 10^6 (Fig. 5.6) or higher.

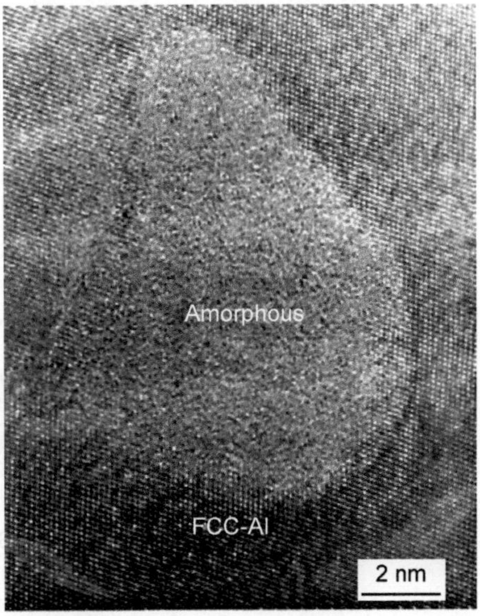

Fig. 5.6 High-resolution TEM image of amorphous nano-pockets in ultrafine-grained Al. (*Source:* BS Murty, IIT Madras).

5.5 ATOMIC FORCE MICROSCOPY (AFM)

AFM is a high-resolution imaging technique that can resolve features as small as an atomic lattice in the real space. It is also commonly referred to as scanning probe microscope (SPM). It allows researchers to both observe as well as manipulate molecular- and atomic-level features. The AFM is being used to solve processing and materials problems in a wide range of technologies affecting the electronics, telecommunications, biological, chemical, automotive, aerospace and energy industries. Almost every material ranging from polymers to ceramics to composites are being investigated using AFM. AFM can be handy in studying the effect of processing/synthesis parameters on the microstructure of the specimen as well as to study the effect of the external environment like chemical or mechanical forces on material behaviour. AFM involves measurement of surface atomic forces in the range of a few nano-Newtons to image the surface topography. Figure 5.7 shows the schematic diagram of an atomic force microscope.

Gerd Binnig and Christoph Gerber invented the AFM in 1985 for which they were awarded the Nobel Prize. The AFM works on the principle of a cantilever, where a small hook is attached to one end of the cantilever and the force between the tip and sample is measured by tracking the deflection of the cantilever as the hook is pressed against the sample surface. This was done by monitoring the tunnelling current to a second tip positioned above the cantilever. Lateral features as small as 30 nm could be seen. The breakthrough in the production of tips/probes has made AFM a reality rather than a curiosity. The first tip prepared was a silicon micro-cantilever, and using it the atomic structure of boron nitride was observed. Today, most of the tips are microfabricated from Si or Si_3N_4. With the imaging of silicon (111) surface, AFM became very popular.

In AFM, a very small force of the order of $<10^{-9}$ N is used between the tip and the sample surface. In AFM, the force is not directly measured, but the deflection of the micro-cantilever or the probe/tip is monitored. The first device introduced by Binnig was

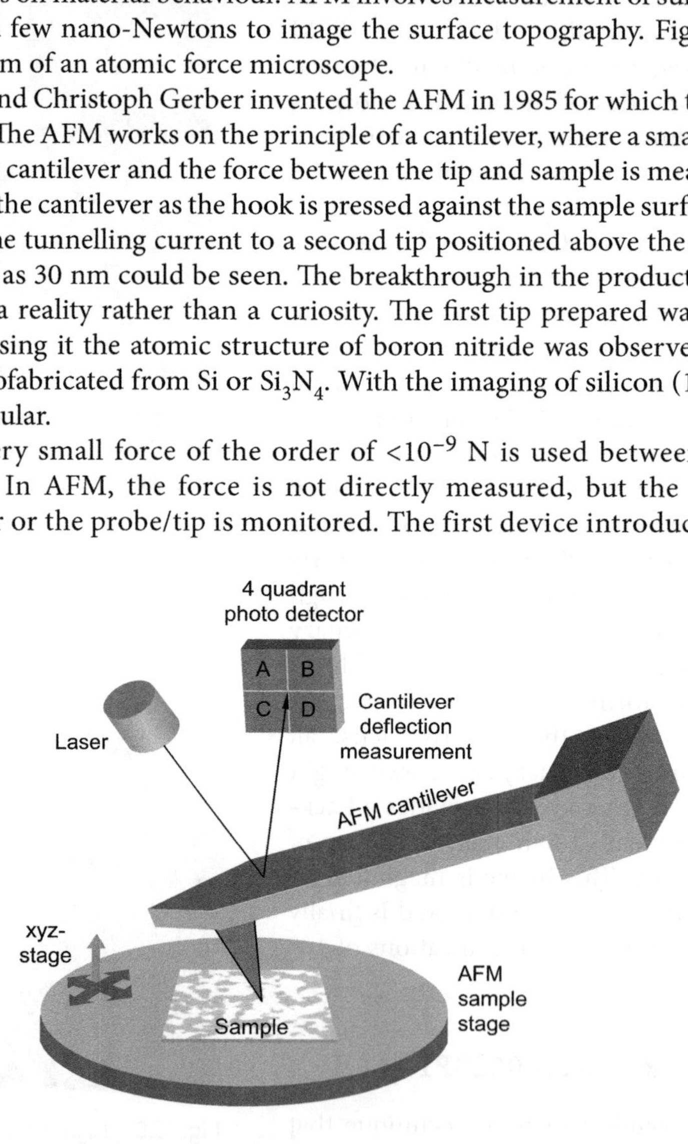

Fig. 5.7 Schematic showing the principle of atomic force microscope (AFM). (*Source:* http://commons.wikimedia.org/wiki/File:AFMsetup.jpg)

a tunnelling tip placed above the metallized surface of the cantilever. This is a sensitive system where a change in spacing of 0.1 nm between the tip and the cantilever changes the tunnelling current by an order of magnitude. It is straightforward to measure deflections smaller than 0.001 nm.

The AFMs that came later are based on optical principles. Small changes in the tilt of the tip can cause changes in optical scattering, which influences the interference pattern and hence the surface can be studied for force variations. In this technique, light is reflected from the surface of the cantilever onto a position-sensitive detector. Thus, even a small deflection of the cantilever will cause a tilt in the reflected light, changing the position of the beam falling on the detector. In another optics-based AFM, the tip acts as one of the mirrors of a diode laser. Any change in the position of the cantilever influences the laser output, which is exploited by the detector. Based on the interaction of the tip with the sample surface, the AFM can be classified as repulsive or contact mode and attractive or non-contact mode. Tapping mode is more widely used.

In the AFM, an atomically sharp tip is scanned over a surface with feedback mechanisms that enable the piezoelectric scanners to maintain the tip at either

1. a constant force (to obtain height information), or
2. at constant height (to obtain force information) above the sample surface.

Tips are typically made from Si_3N_4 or Si, and extended down from the end of a cantilever. In the nanoscope AFM, the tip is attached below a reflective cantilever. A diode laser is focussed on this reflective cantilever. The tip moves on the surface of the sample up and down, tracing the contour of the surface and the laser beam is deflected off the cantilever into a photodiode. The photodetector measures the difference in light intensity between the upper and lower photodetectors, and then converts to voltage. Feedback from the photodiode helps the tip to work in a constant force or a constant height mode. In the constant force mode, the deviation in height is monitored by a piezotransducer. In the constant height mode, the AFM records the deflection force on the sample.

In some AFMs, samples as large as 200 mm wafers can be studied. The primary purpose of these instruments is to quantitatively measure surface roughness with a nominal 5 nm lateral and 0.01 nm vertical resolution on all types of samples. In some AFMs, the sample is translated under the cantilever, while in others the cantilever is moved over the sample. Both these methods can be used to measure the local height of the sample. Plotting of the local sample height as a function of horizontal tip position can give three-dimensional topographical maps of the surface. The concept of resolution in AFM is different from radiation-based microscopy techniques because AFM imaging is a three-dimensional imaging technique. Resolution of the images in optical-based techniques is limited by diffraction, while that in scanning probe techniques is controlled by probe and sample geometry. Usually the width of a DNA molecule is loosely used as a measure of resolution, because it has a known diameter of 2.0 nm in the B form. Some of the best values of resolution for AFM imaging are 3.0 nm based on the DNA in propanol. Unfortunately, this definition of resolution can be misleading because the sample height clearly affects this value.

5.5.1 Comparison of AFM and other imaging techniques

AFM versus STM It is interesting to compare AFM and its precursor, scanning tunnelling microscopy. In some cases, the resolution of STM is better than AFM because of the exponential dependence of the tunnelling current on distance. Only conducting samples can be studied by STM, while AFM can be applied to both conducting and non-conducting samples. AFM is more versatile than STM. In AFM, the voltage and tip-to-substrate spacing can be controlled independently, while in STM these two parameters are connected.

AFM versus SEM Compared with the scanning electron microscope, AFM provides extraordinary topographic contrast, direct height measurements and unobscured views of surface features (no coating is necessary) (Fig. 5.8). Both these techniques measure surface topography. However, both types of microscopes can also measure other surface physical properties. SEM is preferred for measuring chemical composition and AFM for measuring mechanical properties of surfaces.

AFM versus TEM Compared with the transmission electron microscope, three-dimensional AFM images are obtained without expensive sample preparation and yield far more complete information than the two-dimensional profiles available from cross-sectioned samples.

AFM versus optical microscope Compared with the optical interferometric microscope (optical profiles), AFM provides unambiguous measurement of step heights, independent of reflectivity differences between materials.

5.5.2 The common AFM modes

Many AFM modes have appeared for special purposes while the technique of AFM is becoming mature. Three commonly used techniques are discussed here: contact mode, non-contact mode and tapping mode.

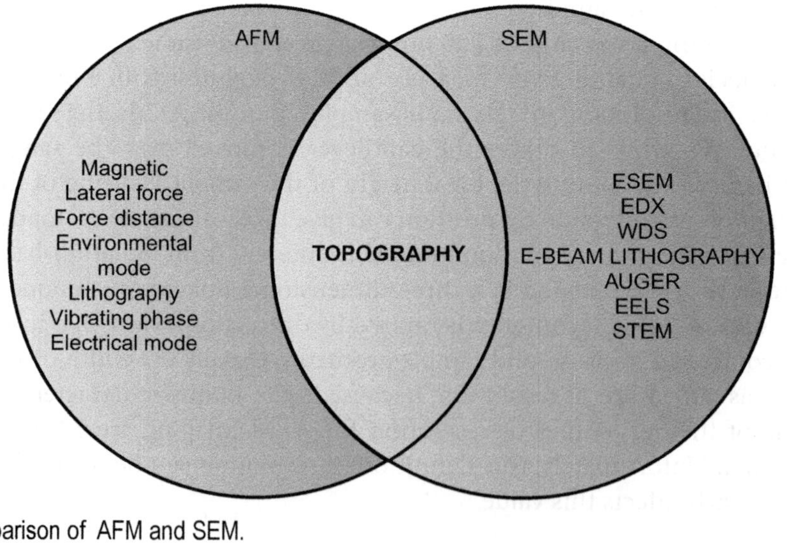

Fig. 5.8 Comparison of AFM and SEM.

Contact mode

Figure 5.9 shows the different contact modes of atomic force microscopy. The most common configuration in contact mode is where the tip is in close contact with the surface being scanned. A repulsive force of about 10^{-9} N is applied on the tip. A piezoelectric positioning element pushes the tip against the sample surface. The deflection of the tip is sensed and compared with a desired value in contact mode AFM. Whenever the measured value does not match the desired value, a voltage is applied to the piezo to change the vertical position of the sample relative to the tip to restore the desired value of deflection. Thus, the voltage applied to the piezo is a measure of the height and depth of features on the surface of the sample. A few instruments operate in UHV but the majority of them operate in ambient atmosphere, or in liquids.

The effect of the impressed force on the surface by the tip can lead to constraints, for example, while studying biological specimens or the effect of adsorbed layers. This can to some extent be minimized by reducing the tracking force of the probe on the sample. Under ambient conditions, sample surfaces are covered by a layer of adsorbed gases consisting primarily of water vapour and nitrogen which is 10–30 monolayers thick. When the tip comes in contact with this contaminant layer, it is pulled by surface tension towards the sample surface. The magnitude of the force depends on the details of the probe geometry, but is typically in the order of 100 nano-Newtons. This meniscus force and other attractive forces may be neutralised by operating with the probe and part or the entire sample stays totally immersed in liquid. Operating the AFM with the sample and cantilever immersed in a fluid has a number of advantages, such as, elimination of capillary forces, reduction of van der Waals forces and liquid–solid interfaces can be studied. However, there could be problems such as liquid leakage, sample damage, etc. All these effects can lead to difficulty in imaging or to destruction of the cantilever probe. Non-contact mode is a preferred technique for imaging such specimens.

Non-contact mode

The non-contact mode of AFM investigation has opened up a new era in imaging by offering far greater versatility in imaging surfaces without fear of damage to surface features or the tip. In this mode, the cantilever tip is placed at a height of about 5–15 nm above the sample surface.

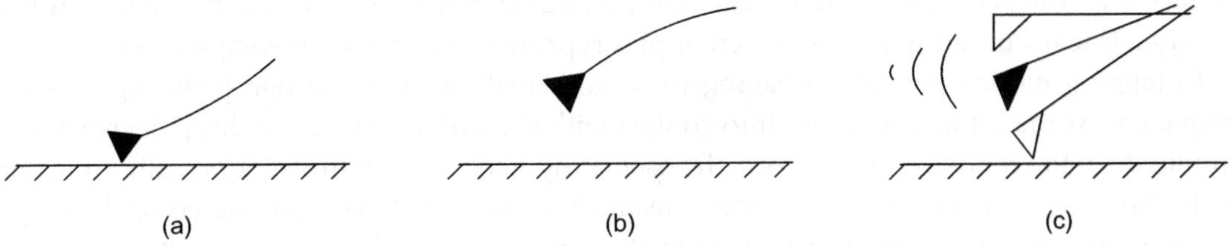

Fig. 5.9 The different contact modes of atomic force microscopy: (a) contact, (b) non-contact and (c) intermittent. (*Source:* http://commons.wikimedia.org/wiki/File:Afm_modes_12-10.png).

The topographic image of the surface is obtained based on the van der Waals forces acting between the tip and the sample. However, these van der Waals forces are weaker than those used in the contact mode. In order to overcome this problem, a small oscillation is given to the tip to detect the small forces between the tip and the sample surface by monitoring the change in amplitude, phase and frequency of the oscillating tip. For highest resolution, it is necessary to measure force gradients from van der Waals forces, which may extend only a nanometre inward from the sample surface. However, as the fluid contaminant layer is substantially thicker than the van der Waals force gradient, the oscillating probe can get trapped in the fluid layer, making imaging with the non-contact tip very difficult.

Tapping mode

The tapping mode helps overcome a few constraints of the contact and non-contact modes of imaging and is a potential technique that allows high-resolution topographic imaging of sample surfaces. The tapping mode overcomes problems associated with friction, adhesion, electrostatic forces and other difficulties that plague conventional AFM scanning methods. In this mode, the tip is oscillated at or near its resonant frequency using a piezoelectric crystal. The piezo makes the tip oscillate with a high amplitude (>20 nm) when the latter is not in contact with the surface. The tip is carefully lowered to bring it in contact with the surface such that it touches it lightly; the oscillating nature of the tip causes the tip to be in contact with the surface and then lift off. The cantilever usually taps the surface at a frequency of about 50,000 to 500,000 cycles per second. Whenever the oscillating tip comes in contact with the surface, its oscillation is reduced due to energy loss. The change in oscillation amplitude is used to identify surface features.

A feedback loop ensures that the cantilever oscillation amplitude is constant during the tapping mode. When the tip passes over a surface protrusion or depression, the amplitude of oscillation decreases or increases, respectively. The digital feedback loop then adjusts the tip–sample separation to maintain constant amplitude and force on the sample.

When the tip is in contact with the surface, the high frequency (50k–500 kHz) makes the surfaces stiff (viscoelastic), and the tip–sample adhesion force is greatly reduced. The tapping mode inherently prevents the tip from sticking to the surface and causing damage during scanning. In the tapping mode, unlike the other two modes, the tip in contact with the surface has sufficient oscillation amplitude to overcome the tip–sample adhesion forces. In addition, as the force is applied vertically the shear forces are negligible and hence the surface material is not pulled sideways. Another advantage of the tapping mode technique is its large, linear operating range, which allows reproducible sample measurements.

In tapping mode operation, imaging in a fluid medium tends to damp the tip's resonant frequency. As the cantilever comes into contact with the surface, there is a sharp decrease in the amplitude of the cantilever oscillation. The oscillating frequency in fluids is normally about 5000 to 40,000 cycles per second. The spring constant is typically 0.1 N/m compared to the tapping mode in air where the cantilever may be in the range of 1–100 N/m. This results in improved imaging in fluids.

The AFM has inspired a variety of other scanning probe techniques. In the earlier part of its development, the AFM was primarily used to understand the topography of surfaces. In recent years, the tip has been modified to measure electric and magnetic properties, chemical potentials, and friction, and even to carry out spectroscopic analysis. The AFM is most often compared with electron beam techniques such as the SEM or TEM. In general, it is easier to learn to use an AFM than an SEM, because there is minimal sample preparation required with an AFM. In AFM, the quality of the image depends to a large extent on the quality of the tip. The following table compares the features of AFM with that of TEM/SEM.

	SEM/TEM	AFM
Samples	Must be conductive	Insulating/Conductive
Magnification	Two-dimensional	Three-dimensional
Environment	Vacuum	Vacuum/Air/Liquid
Time for image	0.1–1 minute	1–5 minute
Horizontal resolution	0.2 nm (TEM) 5 nm (FE-SEM)	0.2 nm
Vertical resolution	n/a	.05 nm
Field of view	100 nm (TEM) 1 mm (SEM)	100 μm
Depth of field	Good	Poor
Contrast on flat samples	Poor	Good

The AFM can easily visualize nanoparticles ranging from a few nanometres to a few microns. It is possible to measure the size of individual nanoparticles as well as measure the parameter distribution of an ensemble of nanoparticles. Parameters such as particle size, volume, circumference and surface area are readily measured. One of the major challenges to measuring nanoparticles is developing methods for distributing the nanoparticles on a surface. Figure 5.10 (see Plate 5) shows the AFM image of a Cu–Ta nanocomposite.

The AFM offers visualization in three dimensions. Resolution in the vertical or z-axis is limited by the vibration environment of the instrument: whereas resolution in the horizontal or xy-axis is limited by the diameter of the tip used for scanning. Typically, AFM instruments have vertical resolutions of less than 0.1 nm and xy-resolutions of around 1 nm.

Software-based image processing of AFM data can generate quantitative information from individual nanoparticles and between groups of nanoparticles. For individual particles, size information (length, width and height) and other physical properties (such as morphology and surface texture) can be measured. Statistics such as particle size distribution on a group of particles can also be measured.

5.6 SCANNING TUNNELLING MICROSCOPE (STM)

The scanning tunnelling microscope (STM) developed by Dr Gerd Binnig and his colleagues in 1981 at the IBM Zurich Research Laboratory, Ruschlikon, Switzerland, is the first instrument capable of directly obtaining three-dimensional (3D) images of solid surfaces with atomic resolution. STMs have been used for the formation of nano-features by localized heating or by inducing chemical reactions under the STM tip and nano-machining. AFMs have been used for nano-fabrication and nano-machining. STMs and AFMs are used at extreme magnifications ranging from 10^3 to 10^9 in the *x*-, *y*- and *z*-directions for imaging macro to atomic dimensions with high resolution information and for spectroscopy. These instruments can be used in any environment such as ambient air, various gases, liquids, vacuum, at low temperatures (lower than about 100 K) and high temperatures.

The principle of *electron tunnelling* was proposed by Giaever. He envisioned that if a potential difference is applied to two metals separated by a thin insulating film, a current will flow because of the ability of electrons to penetrate a potential barrier. Although small, there is a finite probability for tunnelling. To be able to measure a tunnelling current, the two metals must be spaced no more than 10 nm apart. The tunnelling current is highly sensitive to the separation distance between tip and sample; it decreases exponentially with increase in tip–sample separation distance. This factor is crucial in ensuring the excellent vertical resolution of STM (less than 0.1 nm). Tunnelling current decreases by a factor of 2 as the separation is increased by 0.2 nm. Very high lateral resolution depends upon sharp tips. Binnig and his group overcame two key obstacles for damping external vibrations and for moving the tunnelling probe in close proximity to the sample. Their instrument is called the scanning tunnelling microscope (STM). Today's STMs can be used in the ambient environment for atomic-scale images of surfaces.

The principle of the STM is straightforward. A sharp metal tip (one electrode of the tunnel junction) is brought close enough (0.3–1 nm) to the surface to be investigated (the second electrode), such that, at a convenient operating voltage (10 mV–1 V), the tunnelling current varies from 0.2 to 10 nA, which is measurable. The tip is scanned over a surface at a distance of 0.3–1 nm, while the tunnelling current between it and the surface is measured. In STM, a conductive tip placed above the surface of a sample moves on the sample surface with its height being adjusted continuously to keep the tunnelling current constant. The tip position is monitored to map the surface topography of the sample. Figure 5.11 schematically depicts an STM structure. A sharp tip of tungsten or PtIr alloy is mounted on a three-dimensional positioning stage. The tip movement above the sample surface in three dimensions is controlled by piezoelectric arrays. The distance between the tip and the sample is around 0.2 and 0.6 nm, which generates a tunnelling current of about 0.1–l0 nA. The spacial resolution is about 0.01 nm in the *x*- and *y*- directions and about 0.002 nm in the *z*-direction, which leads to a true atomic resolution in three dimensions.

5.6.1 Modes of operation

STM is commonly operated in two modes. One of them is constant current imaging, in which a constant current is maintained between the sample and tip. During the movement of the tip over the sample surface, the vertical position of the tip is changed to maintain a constant separation between the two. Since the tunnelling current is sensitive to distance, constant current imaging will provide excellent surface topographic contrast of the surface atom contours. In contrast to the constant current mode, constant tip position results in variations

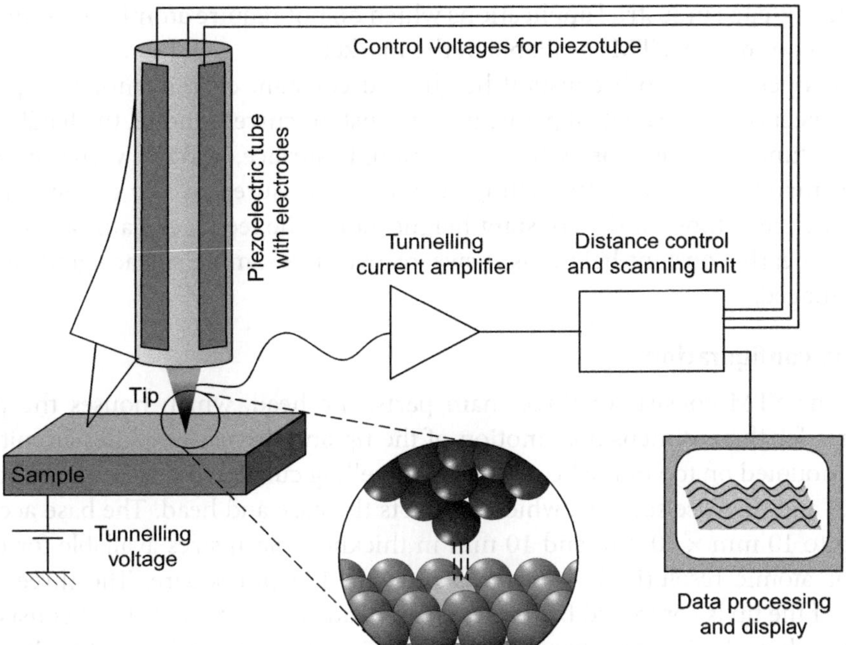

Fig. 5.11 Schematic of scanning tunnelling microscope.

in tunnelling current due to changes in tip separation distance brought about by the 3D topographic features of the surface atoms. In the constant current mode, the contrast is related to electron charge density profiles, while faster scan rates are possible in the constant height mode. STM was first developed by Binnig and Rohrer in 1982, and its atomic-scale resolution was first demonstrated on an image of silicon. After etching the oxide with an HF solution, the (111) silicon wafer was immediately transferred to the STM in a UHV chamber. Repeated heating to 900°C in a vacuum not exceeding 3×10^{-8} Pa resulted in effective sublimation of the SiO layer grown during the transfer, resulting in a clean surface. Only unidirectional scans were recorded to avoid the non-linear effects of the scanning piezoelectric drives.

Families of instruments based on STMs and AFMs, called scanning probe microscopes (SPMs), have been developed for various applications of scientific and industrial interest. These include STM, AFM, FFM (or LFM), scanning electrostatic force microscopy (SEFM), scanning force acoustic microscopy (SFAM) or atomic force acoustic microscopy (AFAM), scanning magnetic microscopy (SMM) or magnetic force microscopy (MFM), scanning near field optical microscopy (SNOM), scanning thermal microscopy (SThM), scanning electrochemical microscopy (SEcM), scanning Kelvin Probe microscopy (SKPM), scanning chemical potential microscopy (SCPM), scanning ion conductance microscopy (SICM), and scanning capacitance microscopy (SCM). Families of instruments that measure forces (e.g., AFM, FFM, SEFM, SFAM and SMM) are also referred to as scanning force microscopy (SFM). Although these instruments offer atomic resolution and are ideal for basic research, they are used for cutting-edge industrial applications which do not require atomic resolution. There are a number of commercial STMs available in the market.

The STM operates in both constant height and constant current modes depending on a parameter selection in the control panel. In the constant current mode, the feedback gains are set high, the tunnelling tip closely tracks the sample surface, and the variation in tip height required to maintain constant tunnelling current is measured by the change in the voltage applied to the piezo tube. In the constant height mode, the feedback gains are set low, the tip remains at a nearly constant height as it sweeps over the sample surface, and the tunnelling current is imaged.

5.6.2 STM configuration

Physically, the STM consists of three main parts: the head, which houses the piezoelectric tube scanner for three-dimensional motion of the tip and the pre-amplifier circuit (FET input amplifier) mounted on top of the head for the tunnelling current, the base on which the sample is mounted, and the base support, which supports the base and head. The base accommodates samples up to 10 mm × 20 mm and 10 mm in thickness. Scan sizes available for the STM are 0.7 μm (for atomic resolution), 12 μm, 75 μm and 125 μm square. The three-dimensional movement of the tip is controlled by the scanning head. The removable head consists of a piezo tube scanner, about 12.7 mm in diameter, mounted into an invar shell used to minimize vertical thermal drifts because of a good thermal match between the piezotube and the invar. The piezo tube has separate electrodes for x, y and z, movement which are driven by separate drive circuits.

The electrode configuration provides x- and y- motions that are perpendicular to each other, minimises horizontal and vertical coupling, and provides good sensitivity. The vertical motion of the tube is controlled by the Z electrode, which is driven by the feedback loop. The x- and y-scanning motions are each controlled by two electrodes, which are driven by voltages of the same magnitude, but opposite signs. These electrodes are called $-Y$, $-X$, $+Y$, and $+X$. The tip holder is a stainless steel tube with a 300 μm inner diameter for 250 μm diameter tips, mounted in ceramic in order to keep the mass on the end of the tube low. The tip is mounted either on the front edge of the tube (to keep mounting mass low and resonant frequency high) or the centre of the tube for large range scanners, namely 75 μm and 125 μm (to preserve the symmetry of the scanning). This commercial STM accepts any tip with a 250 μm diameter shaft. The piezo tube requires X–Y calibration, which is carried out by imaging an appropriate calibration standard. Cleaved graphite is used for the small-scan length head while two-dimensional grids (a gold-plated ruling) can be used for longer range heads.

Characterization of monolayers of nanoparticles using many characterization techniques, like SEM or TEM, is difficult due to the limitations of the resolution offered by the techniques used. Scanning probe techniques have made it possible to investigate the surfaces of materials at the atomic level. In addition, with scanning tunnelling spectroscopy, the electronic structure of surfaces can be studied at molecular level. Though several techniques are used to study such functionalized metal nanoparticles, STM/STS is a unique tool because it enables the topographic and local electronic properties of these films to be investigated with atomic and near-atomic resolution. Two main properties which can be studied with tunnelling spectroscopy on nanoparticles are density of states on isolated nanoparticles and Coulomb blockade and Coulomb staircase phenomena due to charging of the nanoparticles.

5.7 FIELD ION MICROSCOPE (FIM)

Using FIM, individual atoms were observed for the first time. It was developed from its forerunner, the field emission microscope. The specimen prepared in the form of a sharp needle is mounted on an electrically insulated stage. The sample is cooled to cryogenic temperatures in the range of 20–100 K (Fig. 5.12, see Plate 6). The field ion image of the tip of a specimen is observed on a microchannel plate and phosphor screen. An imaging gas is used to produce a field ion image. Neon, helium, hydrogen and argon are the common gases used, and the type of image gas used depends on the material being studied.

The field ion image is produced by the projection of image gas atoms that are ionised by the high positive voltage on the specimen onto the fluorescent screen. The image gas atoms in the vicinity of the specimen are polarized because of the high field and then attracted to the apex region of the specimen. These gas atoms collide with the specimen tip and lose a large part of their kinetic energy. If the field is sufficiently high, these image gas atoms are field ionised by a quantum-mechanical tunnelling process. These ions are repelled from the surface of the specimen towards the microchannel plate and phosphor screen. The image intensifier of the microchannel plate produces about 10^3 to 10^4 electrons for each incident ion. These electrons generate an image on the phosphor screen. Figure 5.13 shows the FIM image of a platinum tip. Each bright spot in the image corresponds to a platinum atom.

A very important area of application of FIM and atom-probe is the micro-analysis of nanomaterials, especially alloys. It is now possible to obtain 3D chemically separated images of nanomaterial structures with atomic resolution and to relate the structure to the mechanical property of these materials. Other important areas of applications are in surface and nanoscience and surface analysis, and also in the development of various field emission sources, including field emission flat panel display and field ionisation and liquid metal ion sources. The major constraint for wide application of FIM is the stringent requirement for sample preparation, need for cryogenic cooling that can cause instability of certain structures, and need for ultrahigh electric field for imaging which limits the material applicability. However, for certain

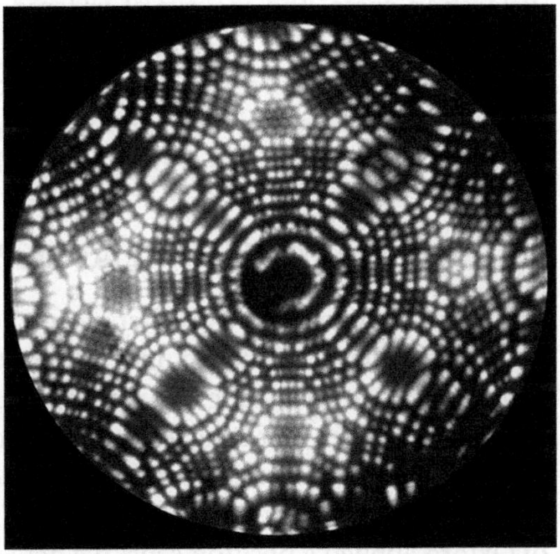

Fig. 5.13 Field ion microscope image of platinum. (*Source:* http://commons.wikimedia.org/wiki/File:FIM-platinum.jpg).

applications, field ion microscopy continues to make unique and valuable contributions that are difficult with other types of atomic resolution microscopy.

Field evaporation is the phenomena by which atoms are ionized in the presence of extremely high electric fields of the order of a few V/nm. Application of a high voltage to a sharp needle-like specimen can generate such high electric fields. Thus, field evaporation can be observed only in field ion microscopy or scanning tunnelling microscopy. Evaporation (ionization) of the atoms start when the electric field is much higher than the field to image FIM. Atom probe measures the time-of-flight of the ions field evaporated from the surface of the FIM tip by controlled application of high frequency pulses.

In the field ion image of a metal surface, individually imaged atoms appear within such a wide range of brightness and size that discrimination between chemically different atom species is usually impossible. With the introduction of the atom probe, field ion microscopy assumes a new dimension by providing the mass-spectrometric identification of a single atom or of the constituents of a small surface area that can be selected from the ion image by the observer. The essential feature of the instrument is the combination of the field ion microscope with the time-of-flight mass spectrometer. The tip is easily exchangeable, can be heated in situ, and can be manipulated to allow any part of the image to fall onto the probe hole. All crystallographic regions are accessible through an external, bellow-sealed ball-bearing mount, while the cold finger, designed to be used with liquid nitrogen, is stationary. The image gas can be supplied dynamically, and the He, Ne, Ar or H ion image is brightly displayed on a 3-inch diameter microchannel ion–electron converter.

The atom probe is a micro-analytical tool of ultimate sensitivity, requiring only one atom for analysis. In metallography it offers itself for the identification of the individual impurity atoms as they are seen in the field ion microscope as interstitials, or segregated to grain boundaries and dislocation cores. The analysis of ordered alloys and precipitation of nuclei is straightforward. Bulk constituents are revealed using the layer-by-layer sectioning of the specimen by field evaporation. Field corrosion products and surface adsorption species can be identified. The specimen is cooled to below 100 K in high vacuum. It is the phenomena occurring at high electric field that are the basis of operation of the FIM and also in the atom probe.

5.8 THREE-DIMENSIONAL ATOM PROBE (3DAP)

Atom probe microscopy was first performed in 1951 by EW Müller at The Pennsylvania State University when he imaged individual atoms with a field ion microscope. He imaged the atomic structure of a tungsten tip and could illustrate that each bright spot corresponds to a single atom. Since that time, enormous progress has been made in instrumentation, especially in signal detection and processing. The most recent advance is the 3D atom probe technique (3DAP), which is now firmly established. In 1988 Cerezo, Godfrey and Smith at the University of Oxford applied a position sensitive detector to a time-of-flight atom probe, thus determining both the mass-to-charge ratio and the position of the ions simultaneously. This new instrument, which they called the position sensitive atom probe (PoSAP), enabled two-dimensional element mapping with sub-nanometre spatial resolution. Reconstruction of a series of two-dimensional maps makes it possible to reconstruct a three-dimensional element map, and hence this new generation of atom probes is generally called three-dimensional atom probe (3DAP).

By measuring the time of flight and the coordinates of ions using a PSD, it is possible to map out two-dimensional elemental distributions with near-atomic resolution. The lateral spatial resolution is limited by the evaporation aberration that occurs during the ionization of atoms on the surface. However, the error originating from the evaporation aberration does not exceed 0.2 nm, and this is still the lowest of the errors achieved by existing analytical instruments. The elemental maps can be extended to the depth direction by ionising atoms from the surface of the specimen continuously, by which atom distribution can be reconstructed in a three-dimensional real space. Since field evaporation occurs layer-by-layer in low index planes, the reconstructed 3D elemental map shows the layers corresponding to the atomic planes in the depth direction. Currently, the atom probe is used routinely to examine materials literally on an atom-by-atom basis.

3DAP allows the analysis of cylindrical parts of a specimen (with a diameter of about 30 nm and a length of about 200 nm), probing atom by atom through the specimen. The technique has enormous potential for understanding the fundamental aspects of materials science, including segregation at grain boundaries, processes of nucleation in materials, radiation-induced defects, and decomposition of materials. Impressive results were also obtained on the thermal stability and interdiffusion in thin films. An example of the ultrafine-scale information available with the 3DAP technique is shown in Fig. 5.14, which represents the atomic imaging of a Ω-phase precipitate formed in an Al alloy. Segregation to the surface of the precipitate can easily be recognized. The example illustrates the detailed, ultrahigh-resolution chemical information that is available with an advanced 3DAP. Figure 5.14 also shows the elemental

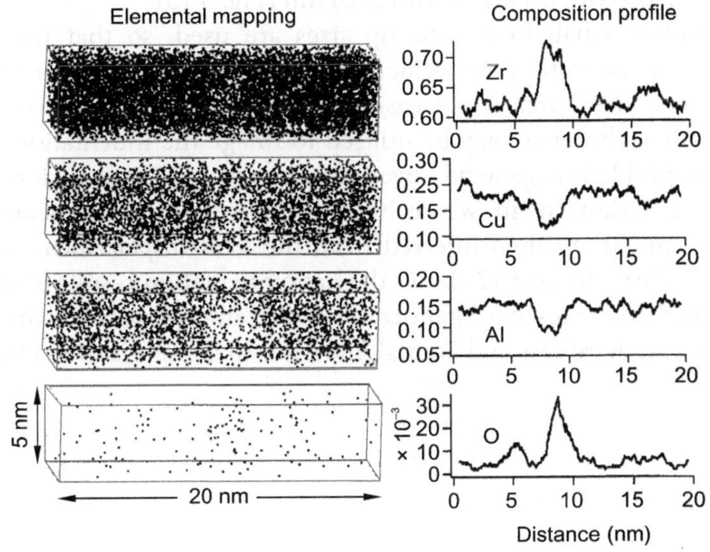

Fig. 5.14 Elemental maps obtained in 3DAP from a Zr alloy. (*Source:* BS Murty, IIT Madras).

maps of Zr, Cu, Al and O obtained from 3DAP of a Zr alloy, which shows oxygen segregation into a quasicrystalline particle. This technique could establish that oxygen stabilises the nano-quasicrystalline phase.

5.9 NANOINDENTATION

Nanoindentation is a relatively new technique to obtain mechanical properties of nanometric regions by studying the stress–strain behaviour when a nano-indentor is impressed against the specimen of interest. Usually scanning force microscope (SFM) tips are used both to perform the indentation and to image the sample surface after indentation. The depth of indentation is measured as a function of the force (stress) applied. The mechanical properties including elastic constants can be derived from such loading–unloading curves. A diamond indenter with a small radius of curvature at its tip is generally used. This nanoindentation device can be mounted on the scanner head of the AFM in place of the cantilever. The principle of nanoindentation is similar to other indentation hardness testers, in which an indentor is forced into the material being tested, forming an indent. The hardness is taken as the ratio of load and the area of contact between the sample and the indenter. In nanoindentation, the contact area is measured by a depth-sensing technique. During the test, both the load and the displacement of the indenter are recorded. The load–displacement curve is used to calculate the contact area at maximum load. This presents the problems related to the measurement of extremely small contact areas.

With the help of a nano-indentor, one can measure the elastic modulus also, in addition to hardness. These two properties are a measure of resistance offered by the material to elastic and plastic deformation, respectively. The use of the AFM along with nanoindentation helps in positioning of the tip with accuracy within 20 nm (Fig. 5.15).

In nanoindentation, small loads and tip sizes are used, so that the indentation area may measure only a few square micrometres or even nanometres. Due to this, a typical problem is that the contact area is not easily found. Atomic force microscopy or scanning electron microscopy techniques may be utilised to image the indentation, but can be quite cumbersome. This problem is generally overcome by using high-precision geometry for the indentor, namely, a Berkovich tip, which has a three-sided pyramid geometry. Mechanical properties of the sample are then inferred from the *load–displacement curve* (such as the one shown in Fig. 5.16). Figure 5.17 shows the nanoindentation results of different phases in the Zr–Pt alloy, which indicates that a nanocomposite of amorphous and nano-quasicrystalline phase has much higher hardness and Young's modulus in comparison to individual phases.

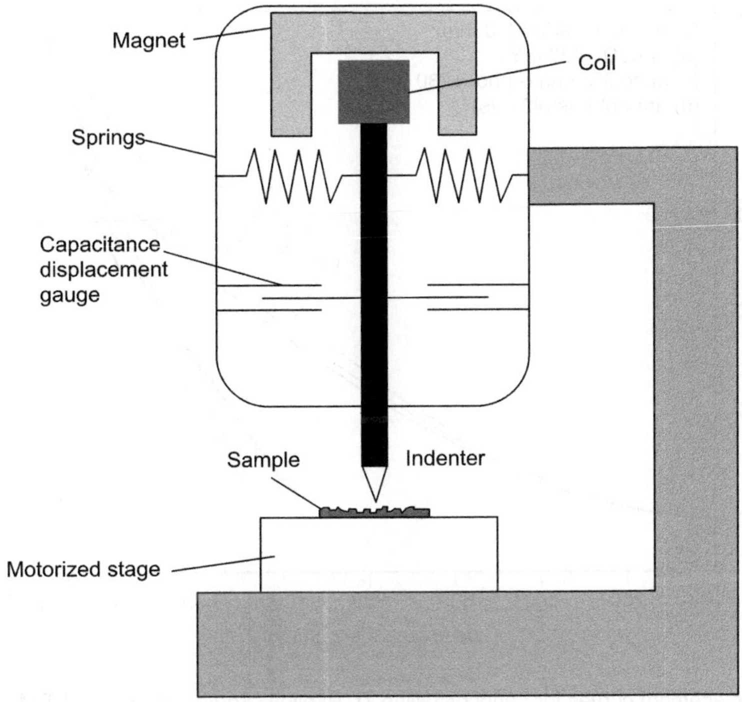

Fig. 5.15 Schematic diagram of nanoindentation mechanism.

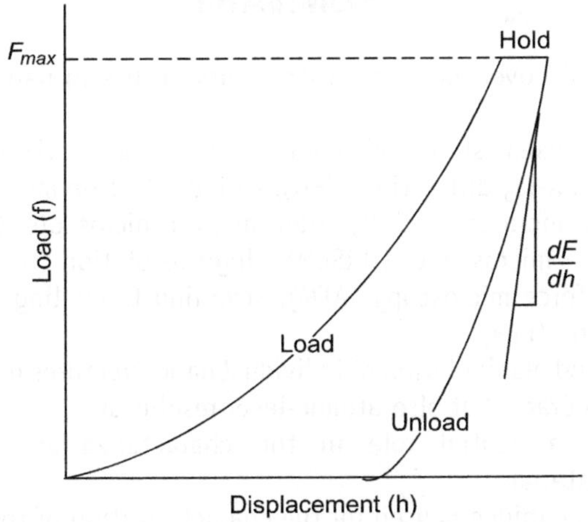

Fig. 5.16 Load–displacement curve of nanoindentation technique (*Source:* http://commons.wikimedia.org/wiki/File:Load-displacement_curve_%2B_onderdelen.JPG).

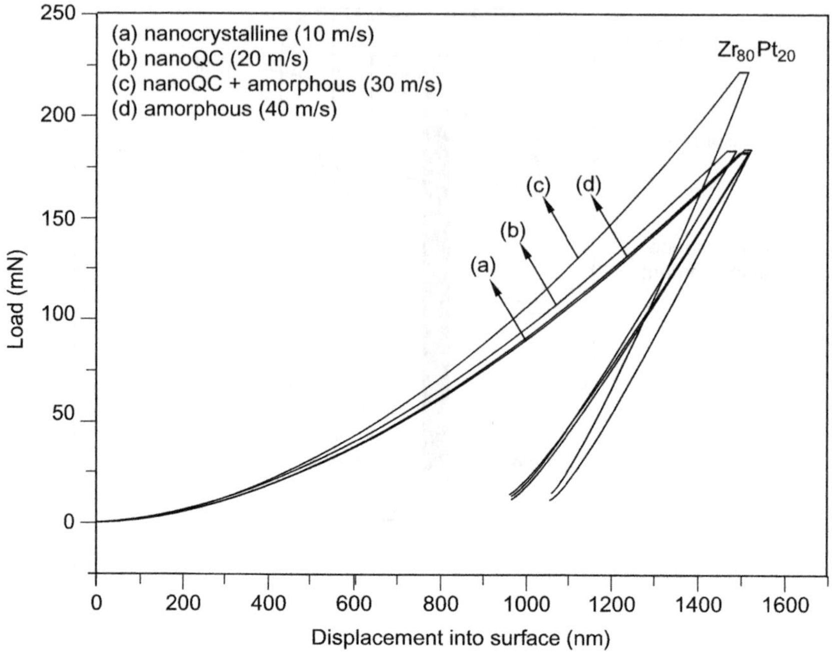

Fig. 5.17 Load–displacement curves for nanocrystalline Zr–Pt alloy (*Source:* BS Murty, IIT Madras).

SUMMARY

- The development of novel tools and instruments for use in nanotechnology is a great challenge.
- The most widely used structural methods for characterizing nanomaterials and nanostructures are: X-ray diffraction (XRD), various electron microscopy (EM), including scanning electron microscopy (SEM), transmission microscopy (TEM), high-resolution scanning electron microscopy (HRSEM), high-resolution transmission microscopy (HRTEM), atomic force microscopy (AFM), scanning tunnelling microscopy (STM) and field ion microscopy (FIM).
- Characterization and manipulation of individual nanostructures require not only extreme sensitivity and accuracy, but also atomic-level resolution.
- Microscopy plays a central role in the characterization and measurement of nanostructured materials.
- Nanoindentation technique is used for the characterization of the mechanical property of nanomaterials by studying the force–displacement curve on application of very small loads.

EXERCISES

1. Construct the spotty ring selected area electron diffraction pattern for the rutile and anatase phase of nano-titania, considering that TEM is operated at 120 keV and the diffraction pattern is recorded at an appropriate camera length. Describe the expected changes in the electron diffraction pattern as the crystallite size reduces from 50 nm to 10 nm and finally to 5 nm.
2. Generate a schematic of the expected X-ray diffraction pattern (CuK-α) from an FCC crystalline substrate with a crystallite size of 20 nm (considering spherical grains). Calculate the increase in peak broadening expected as the crystallite size reduces from 20 nm to 2 nm.
3. Explain in brief the bright field, dark field and phase contrast modes of operation in a TEM.
4. Plot a typical force–displacement curve obtained from a nanoindentation loading–unloading curve. Explain how the mechanical properties of the material can be inferred from this curve.
5. Describe the different operating modes in an AFM. Compare and contrast between AFM and STM.

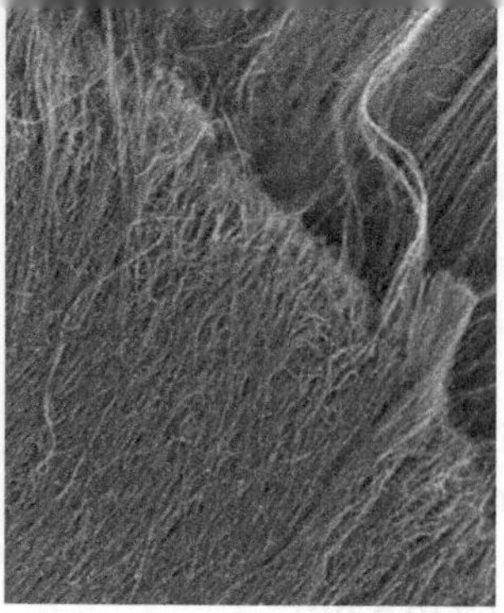

Chapter 6

Nanostructured Materials with High Application Potential

Learning objectives

- Different kinds of nanomaterials that have shown immense potential for applications:
  - Fabrication and applications of quantum dots
  - Synthesis, characterization and applications of single-walled and multi-walled carbon nanotubes
  - Synthesis and applications of GaN nanowires
  - Synthesis and applications of nanocrystalline ZnO
  - Synthesis and applications of nanocrystalline TiO_2

Materials development has remained the backbone of human civilization and will continue to be the anchor for all future developments. The development of new materials and advanced material technologies has served as the cradle for most engineering developments. Revolutions in the communication, computing, energy, chemical, transport and engineering industries have been possible only due to credible advances in materials technology and the development of new classes of materials. As discussed in earlier chapters, nanomaterials are slowly beginning to make their presence felt in science and technology. The potential engineering applications of nanomaterials are vast. This chapter will describe only a few typical nanostructured materials of current interest. One extreme end of nanostructures is single electron transistors. Figure 6.1 shows an example of a single electron transistor with niobium leads and aluminium islands.

6.1 QUANTUM DOTS

The research in microelectronic materials is driven by the need to tailor electronic and optical properties for specific component applications. Progress in epitaxial growth, advances in patterning and other processing techniques have made it possible to fabricate 'artificial' dedicated materials for microelectronics. In these materials, the electronic structure is tailored by changing the local material composition and by confining the electrons in nanometre-size foils or grains. Due to quantization of electron energies, these systems are often called *quantum structures*. If the electrons are confined by a potential barrier in all three directions, the nanocrystals are called quantum dots (QDs). QDs have unique properties that fall between

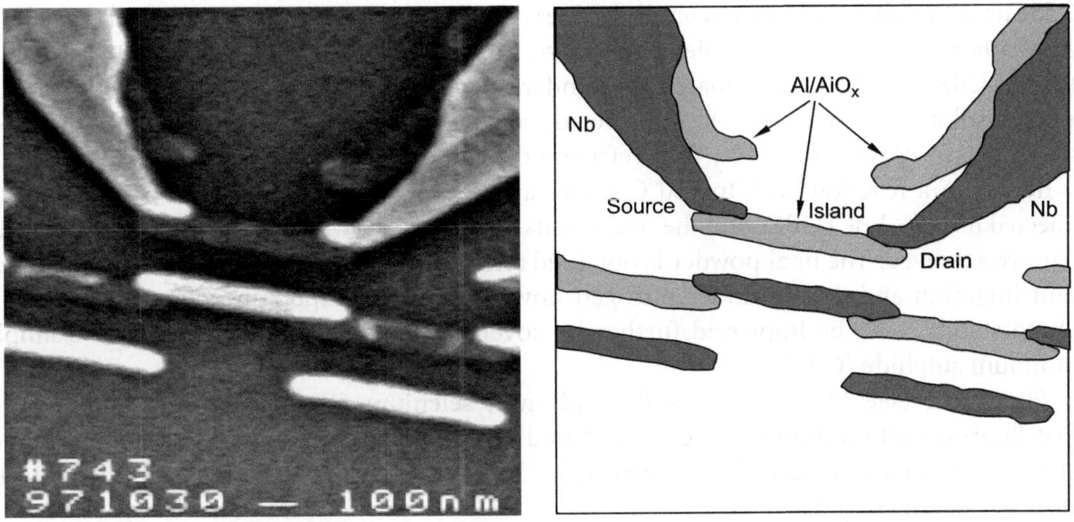

(a) (b)

Fig. 6.1 Single electron transistor with niobium leads and aluminium islands: (a) scanning electron micrograph, and (b) artistic interpretation. (*Source:* http://commons.wikimedia.org/wiki/File:TySETimage.png).

bulk semiconductors and individual molecules. QDs in transistors, solar cells, LEDs, diode lasers, etc., have been studied. They have also been used for medical imaging and there is hope that they can be used as *qubits**. Some QDs are commercially available.

6.1.1 Fabrication

There are several ways to confine excitons in semiconductors, resulting in different methods of producing QDs. In the following section, the discussion is limited to selected promising QD technologies including semiconductor nanocrystal QDs (NCQD), lithographically made QDs (LGQD), field effect QDs (FEQD) and self-assembled QDs (SAQD).

Semiconductor nanocrystal QDs (NCQD)

A nanocrystal (NC) is a single crystal with a diameter of a few nanometres. An NCQD is a nanocrystal which has a smaller band gap than the surrounding material. Grinding of a macroscopic crystal can produce NCQDs. NCQDs are attractive for optical applications because their colour is directly determined by their dimensions. The size of the NCQD can be selected by filtering a larger collection NCQD or by tuning the parameters of a chemical fabrication process.

---

*A quantum bit or *qubit* is a unit of quantum information. A bit is the basic unit of computer information. It takes a value of either 0 or 1. A qubit is similar to a bit. Like a bit, a qubit can be 0 or 1. The difference is that whereas a bit *must* be either 0 or 1, a qubit can be 0, 1, or a superposition of both.

CdSe nanocrystals Cadmium selenide (CdSe) NCQDs are approximately spherical crystallites with either wurtzite or a zinc-blend structure. Their diameter usually ranges between 1 and 10 nm. CdSe NCQDs are prepared by standard processing methods. $Cd(CH_3)_2$ is added to a stock solution of selenium (Se) powder dissolved in tributylphosphine (TBP). This stock solution is prepared under N_2 in a refrigerator, while tri-*n*-octylphosphine oxide (TOPO) is heated in a reaction flask to 360°C under argon flow. The stock solution is then quickly injected into the hot TOPO, and the reaction flask is cooled down when NCQDs of the desired size are achieved. The final powder is obtained after precipitating the NCQDs with methanol, centrifugation and drying under nitrogen flow. The room temperature quantum yield and photo stability can be improved further by covering the CdSe NCQDs with, for example, cadmium sulphide (CdS).

Figure 6.2 (see Plate 6) shows the cadmium selenium QDs, which are nanoparticles that fluoresce into a number of colours based on their size. In these QDs, a small number of electrons are confined within a small space by the placement of some insulating material around a conducting material.

Silicon nanocrystals Silicon/silicon dioxide (Si/SiO_2) NCQDs are Si clusters completely embedded in insulating SiO_2. They are fabricated by ion-implanting Si atoms into either ultra-pure quartz or thermally grown SiO_2. The NCQDs are then formed from the implanted atoms under thermal annealing. The exact structure of the resulting NCQDs is not known. Scientists have successfully fabricated NCQDs with a diameter around 3 nm and an NCQD density of 2×10^{-19} cm^{-3}. The high density results in even higher light wave amplification (100 cm^{-1}) than for 7 stacks of InAs QDs (70–85 cm^{-1}). The main photoluminescence peak was measured at $\lambda = 800$ nm. The radiative recombination in these QDs is not very well understood, though it has been suggested that radiative recombinations take place through interface states. Despite the very high modal gain, it is very difficult to fabricate an electrically pumped laser structure of Si NCQD due to the insulating SiO_2 matrix.

Lithographically made QDs (LGQD)

A vertical QD (VQD) is formed by etching out a pillar from a quantum wall (QW) or a double barrier heterostructure (DBH). Figure 6.3 shows the main steps in the fabrication process of a VQD. The Al-GaAs/InGaAs/AlGaAs DBH was grown epitaxially, after which a cylindrical pillar was etched through the DBH. Finally, metallic contacts were made for electrical control of the QD. Typical QD dimensions comprise a diameter of about 500 nm and thickness of about 50 nm. The confinement potential due to the Al-GaAs barriers is about 200 meV. The optical quality of VQDs is usually fairly poor due to the etched boundaries. However, VQDs are attractive for electrical devices because of the well-controlled geometry and the well-defined electrical contacts.

Field effect quantum dots (FEQD)

In a field effect QD (FEQD), the charge carriers are confined into a 2D electron gas (2DEG) by a modulation-doped heterojunction. Within the 2DEG plane, the charges are electrostatically confined by external gates. Figure 6.4 shows the schematic of typical device geometry. The

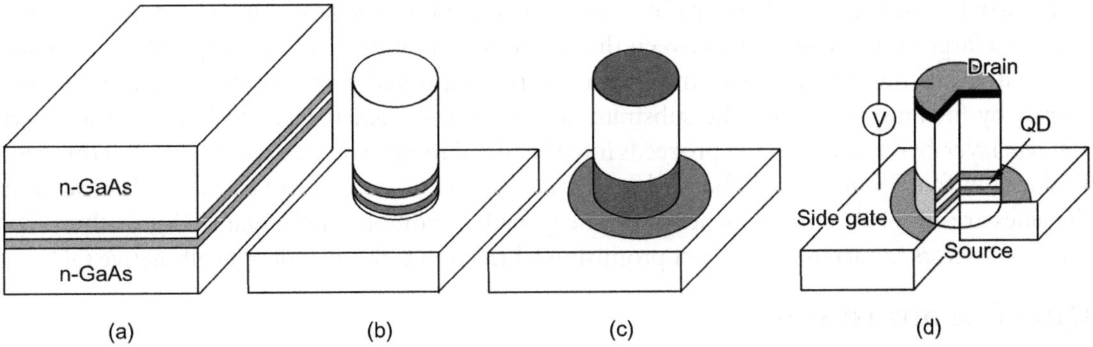

Fig. 6.3 Schematic representation of the fabrication process of a VQD. (a) Epitaxial growth of a DBH, (b) etching of a pillar through the DBH and (c) the metallization.

Ohmic contacts in Fig. 6.4 represent any kind of electric contact to the QD. The effective potential of an FEQD is very smooth and, within the plane of the 2DEG, its shape is close to a parabola, depending on the gates. For an FEQD with a diameter around 200 nm, the spacing of the energy level is typically tens of μeV. These types of QDs are not expected to operate at room temperature because of the shallow potential profile. However, FEQDs are attractive for low temperature infrared light detectors because of a very smooth gate-induced potential and high-quality heterostructure interfaces.

Self-assembled quantum dots (SAQD)

In self-assembly of QDs, one makes use of an island formation in epitaxial growth. This is similar to forming water droplets on a polished surface. These islands are either QDs themselves or QDs formed on a QW. The major self-assembly growth techniques are vapour phase epitaxy (VPE) and molecular beam epitaxy (MBE).

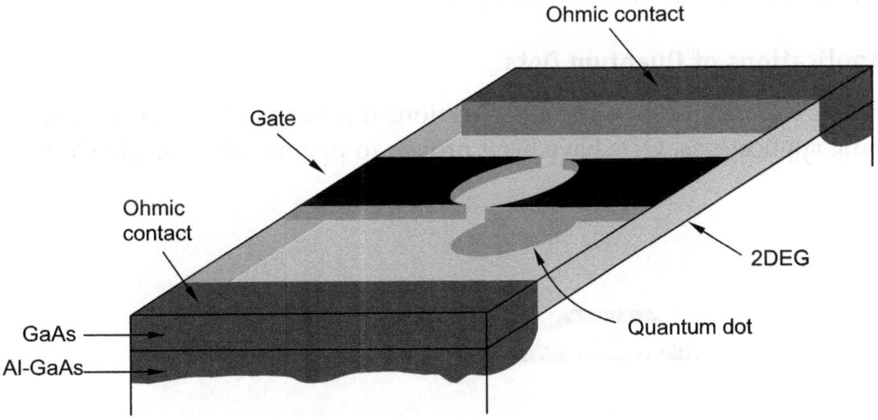

Fig. 6.4 A schematic drawing of an FEQD in a 2DEG at the material interface between Al-GaAs and GaAs.

Epitaxial growth usually happens by layered growth of atoms. However, islands are formed if there is a large lattice mismatch between the materials, and if the surface energy of the deposited material is different from that of the substrate. The deposited material minimizes its potential energy by forming islands on the substrate. In the Stranski–Krastanow (SK) mode, the growth starts in layer-by-layer mode and proceeds into the island mode when it exceeds a critical thickness. Dislocation free SK growth has been observed in InAs on GaAs and InP on GaAs. Typical island densities are 10^9–10^{12} cm^{-2} depending on the growth conditions. Self-organized growth of III-V semiconductors is currently the most promising fabrication technique of optically active QDs.

Quantum Dot Island

The self-assembled island is in itself a QD, if the island is embedded in a material with a larger band gap than that of the island material. An example is provided by InAs islands in GaAs. Figure 6.5 shows QD islands schematically and a high-resolution scanning tunnelling micrograph of a true InAs island. Very promising laser structures have been fabricated using these types of QDs by stacking several island layers on top of each other. Typical heights of QDs range from 5 to 15 nm and widths from 15 to 25 nm. Each QD has very few electrons and holes. The total charge confinement is a combination of strain, piezoelectric fields and material interface effects. For a dot of 13.6 nm height, the calculated confinement energy of the electron ground state is about 180 meV.

Strain-Induced Quantum Dots

Strain is always present in self-assembled islands as well as in the substrate close to the island. The magnitude of the strain depends on the lattice constants and elastic moduli of the materials. If there is a QW close to the QD, the strain field penetrates it and affects its energy bands. The QD can induce a lateral carrier confinement in QW, leading to a total QD confinement in QW. Typical stressor island heights range from 12 to 18 nm and the QW thickness is around 10 nm. The lateral strain-induced confinement is very smooth and has the shape of a parabola. The strain-induced electron confinement is about 70 meV deep. The resulting QD is large and contains, in general, tens of electron-hole pairs.

6.1.2 Applications of Quantum Dots

Quantum dots can be used in optical applications due to their large possible quantum yield. In electronic applications, QDs have been proven to operate like a single electron transistor

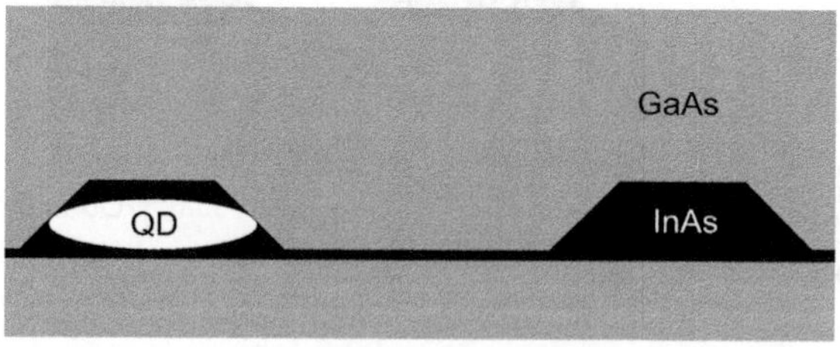

Fig. 6.5 Schematic image of an InAs QD island embedded in GaAs.

and show the Coulomb blockade effect. QDs are also being suggested as qubits for quantum information processing. The main advantage of QDs is the ability to tune their size. Larger QDs show red shift in their spectra in comparison to smaller QDs. The smaller QDs show significant quantum properties. As QDs are zero-dimensional structures, they have sharper density of states in comparison to higher-dimensional structures. QDs have excellent transport and optical properties, and are hence used in diode lasers, amplifiers and biological sensors.

Computing

QD technology is one of the most promising candidates for use in solid-state quantum computation. The application of a small voltage can control the flow of electrons through the QD and thereby make precise measurements of the spin and other properties therein. With the help of several entangled QDs, quantum calculations might be possible.

Biology

A number of organic dyes are used in biological analysis. However, more flexibility is being required of these dyes, and the traditional dyes are often unable to meet the expectations. QDs have been found to be a good replacement for these traditional dyes as they offer better brightness and stability. QDs are not suitable for single-particle tracking due to their irregular blinking. Semiconductor QDs have also been used for in vitro imaging of pre-labelled cells. QDs are being used to study embryogenesis, cancer metastasis, stem-cell therapeutics and lymphocyte immunology due to their ability to image single-cell migration in real time.

Scientists have proved that QDs are dramatically better than existing methods for delivering a gene-silencing tool, known as SIRNA, into cells. Attempts have been made in using QDs for tumour targetting under in vivo conditions. Active and passive targetting are the two types of tumour targetting employed. In the case of active targetting, QDs are functionalized with tumour-specific binding sites to specifically bind to tumour cells. Passive targetting utilises enhanced permeation and retention of tumour cells for the delivery of QD probes. Nanoparticles can easily enter into the fast growing tumour cells as they have more permeable membranes than healthy cells. Nanoparticles can accumulate in tumour cells as they lack an effective lymphatic drainage system.

One of the remaining issues with QD probes is their in vivo toxicity. CdSe nanocrystals, for example, are highly toxic to cultured cells under UV illumination. The energy of UV radiation is close to the covalent chemical bond energy of CdSe nanocrystals. CdSe particles can be dissolved in a culture medium, releasing toxic cadmium ions. In the absence of UV irradiation, however, QDs with a stable polymer coating have been found to be essentially non-toxic. Little is known about the excretion process of polymer-protected QDs from living organisms. These and other questions must be carefully examined before QD applications in tumour or vascular imaging can be approved for human clinical use. Another cutting-edge application of QDs as potential artificial fluorophores for intra-operative detection of tumours using fluorescence spectroscopy is also being researched.

Photovoltaic devices

Quantum dots may have the potential to increase the efficiency and reduce the cost of today's typical silicon photovoltaic cells. According to experimental proof from 2006, QDs of lead selenide can produce as many as seven excitons from one high-energy photon of sunlight (7.8 times the band gap energy). This compares favourably to current photovoltaic cells which can only manage one exciton per high-energy photon, with high kinetic energy carriers losing their energy as heat. This would not result in a 7-fold increase in final output, however, but could boost the maximum theoretical efficiency from 31% to 42%. QD photovoltaics would theoretically be cheaper to manufacture, as they can be made using simple chemical reactions.

Light emitting devices

There are several inquiries into using QDs as light-emitting diodes to make displays and other light sources: QD–LED displays and QD–WLED (White LED). In June 2006, QD Vision announced technical success in making a proof-of-concept QD display. QDs are valued for displays, because they emit light in very specific Gaussian distributions. The display obtained by QDs can show colours more accurately than the human eye can perceive. QDs also require very little power since they are not colour filtered. Additionally, since the discovery of 'white-light emitting' QD, general solid-state lighting applications are closer than ever. A liquid crystal display (LCD), for example, is powered by a single fluorescent lamp that is colour-filtered to produce red, green and blue pixels. The efficiency of displays with monochromatic light is higher as more of the light reaches the eye.

6.2 CARBON NANOTUBES

Carbon is a unique element. It can exist in several different allotropic forms at room temperature, namely graphite, diamond, amorphous carbon, carbon clusters (like C_{60}, C_{70}, etc.) and carbon nanotubes. From the soft graphite to the hard diamonds, the electrically conducting graphite to insulating diamonds and semiconducting CNTs, lustrous diamond to opaque graphite, carbon can exhibit extreme variations in a given material property through its various structural forms. Although diamond, graphite and CNTs are compositionally made of the same matter, they exhibit extreme variation in properties due to differences in their bonding and structure. This aspect in particular makes the study of carbon materials scientifically exciting.

Carbon atoms can be chemically bonded to each other either in the sp^2 (graphite, carbon clusters, CNT) or sp^3 (diamond) hybridized state. In diamond-like carbon (DLC) films, there is a random network of the two hybridized states. In fullerenes (C_{60}, or 'bucky ball') the carbon atoms are sp^2 hybridized, but in contrast to graphite, they are not arranged on a plane. The geometry of C_{60} strains the bonds of the sp^2 hybridized carbon atoms, creating new properties for C_{60}. Graphite is a semi-metal, whereas C_{60} is a semiconductor. Fullerenes were discovered in 1985 by Rick Smalley and co-workers. C_{60} was the first fullerene prepared. It is a football-shaped (icosahedral) molecule with 60 carbon atoms bonded together in pentagons and hexagons (Fig. 6.6). Fullerenes get their name from Buckminster Fuller, an architect who built a dome that has the same structure as that of the C_{60} molecule. This is possibly the first time that an architect's name was associated with a scientific discovery.

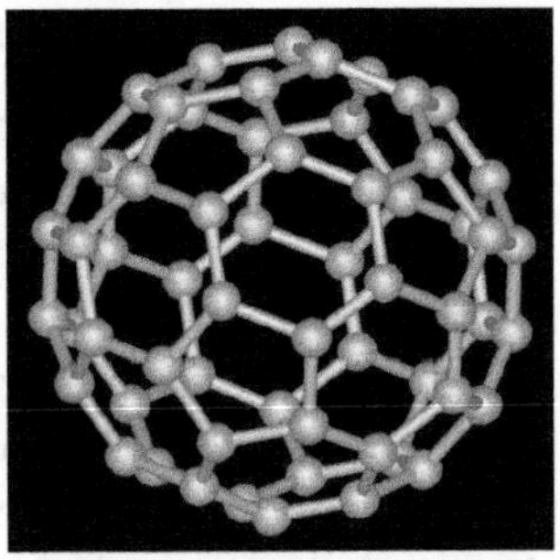

Fig. 6.6 Fullerene C_{60} ball and stick created from a PDB using Piotr Rotkiewicz's source (*Source:* original upload2cebway2005 by IMEOW bot category: Fullerenes).

6.2.1 Types of carbon nanotubes

A carbon nanotube is a planar sheet of graphite (called graphene) rolled up into a seamless cylinder with diameter in the order of a nanometre. This results in a nanostructure where the length-to-diameter ratio exceeds nearly 10,000. Each end of the long cylinder is capped with half a fullerene molecule.

CNTs can in general be classified as either single-walled nanotubes (SWNTs) or multi-walled nanotubes (MWNTs). SWNTs have a cylindrical shell with one atom thickness. The concentric arrangement of several single-walled nanotubes of slightly varying diameters is termed as a multi-walled nanotube. Iijima was the first to recognise that nanotubes were concentrically rolled graphene sheets with a large number of potential helicities and chiralities rather than a graphene sheet rolled up like a scroll, as originally proposed by Bacon (Fig. 6.7; see Plate 7). Carbon nanotubes are unique nanostructures with remarkable electronic and mechanical properties, some of which are due to the close relation between the carbon nanotubes and graphite, and some to their one-dimensional aspect.

Single-walled nanotubes

Most single-walled nanotubes (SWNT) have diameter close to 1 nanometre, with a tube length that can be many thousands of times longer. Single-walled nanotubes with length up to orders of centimetres have been produced. Single-walled nanotubes are a very important variety of carbon nanotube because they exhibit important electric properties that are different from those of multi-walled carbon nanotube (MWNT) variants. It is expected that single-walled nanotubes can help in further miniaturising electronics beyond MEMS. SWNTs is the first intramolecular field effect transistor (FETs) to be synthesized. Single-walled nanotubes are still very expensive to produce, and the development of more affordable synthesis techniques is vital to the future of carbon nanotechnology.

Multi-walled nanotubes

Multi-walled nanotubes (MWNT) consist of multiple layers of graphite rolled in on themselves to form a tube shape. The distance between these layers is close to the graphene layer distance in graphite. This is especially important when functionalisation is required to add new properties to CNTs. Functionalisation will break some C=C double bonds on the SWNTs, which can modify their mechanical and electrical properties.

6.2.2 Chirality of carbon nanotubes

Nanotubes can also be classified as armchair nanotubes, zigzag nanotubes or chiral tubes depending on their chirality. *Chirality* can be described by the chiral vector (n, m), where n and m are integers of the vector equation $\mathbf{R} = n\mathbf{a}_1 + m\mathbf{a}_2$; where $\mathbf{a}_1$ and $\mathbf{a}_2$ are the basis vectors of the hexagonal 2D lattice. **R** is the rolling vector about which the planar graphite sheet (of width equal to the perimeter of the nanotube along the cross section) is rolled to form the CNT. The armchair vector is defined as the vector from a given atom position dividing the hexagons into half. If the armchair vector coincides with the rolling vector (wrapping angle, i.e., angle between the armchair vector and rolling vector, is zero), it results in an armchair nanotube. In other words, the chiral vector **R** will be given as $\mathbf{R} = n\ (\mathbf{a}_1 + \mathbf{a}_2)$ (i.e., $n = m$ in the chiral vector). If the wrapping angle is equal to 30° ($\mathbf{R} = n\mathbf{a}_1$, with m being zero), then the tube is of the 'zigzag' type. If the wrapping angle is between zero and 30°, it is called a chiral tube with $\mathbf{R} = n\mathbf{a}_1 + m\mathbf{a}_2$; with $n \neq m$. Consider a nanotube that has been unfolded into a planar sheet by splitting along a circumferential vector parallel to the tube axis. The values of n and m determine the chirality, or 'twist' of the nanotube. The chirality in turn affects the conductance of the nanotube, its density, lattice structure and other properties. The value of $n - m$ decides whether SWNT is metallic or semiconducting—metallic if this value is divisible by three, semiconducting otherwise.

6.2.3 Synthesis of CNT

Electric arc discharge

In arc-discharge, carbon atoms are evaporated by the plasma of helium gas ignited by high currents passed through the opposing carbon anode and cathode (Fig. 6.8). Arc discharge synthesis uses a low-voltage (~12 to 25 V), high-current (50 to 120 A) power supply (an arc welder can be used). An arc is produced across a 1-mm gap between two graphite electrodes 5 to 20 mm in diameter. An inert gas such as He or Ar is used as the atmosphere for the reaction, at a pressure of 100 to 1000 torr. The arc discharge method involves condensation of carbon atoms generated from the evaporation of solid carbon atoms. The temperature involved with this method is close to the melting point of graphite, 3000–4000°C. Arc discharge has been developed into an excellent method for producing high-quality multi- and single-walled nanotubes.

MWNTs can be grown by controlling growth conditions such as the pressure of inert gas in the discharge chamber and the arcing current. MWNTs synthesized previously by this method have lengths in the order of ten microns and diameters in the range of 5–30 nm. Individual nanotubes form bundles with other nanotubes, which are all held together by van der Waals interactions. MWNTs produced by arc discharge are very straight which indicates their high crystallinity. The growth of single-walled nanotubes requires a metal catalyst in the arc discharge system. The nanotubes produced are mostly in the form of ropes consisting of tens of individual nanotubes close-packed into hexagonal crystals via van der Waals interactions. Arc discharge was the first recognized method for producing both SWCNTs and MWCNTs, and has been optimized to be able to produce gram quantities of either.

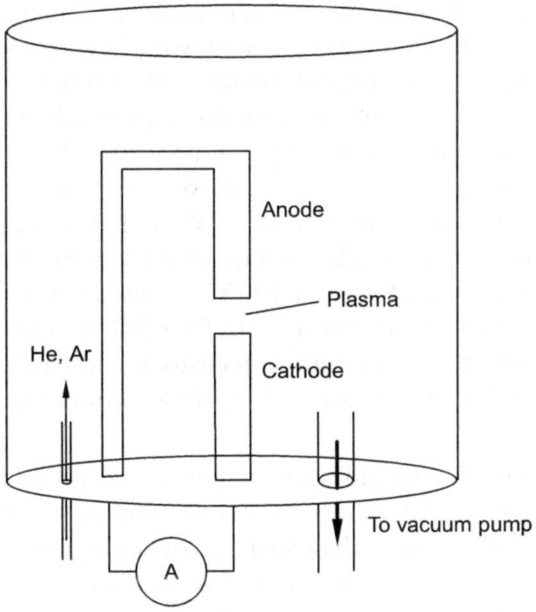

Fig. 6.8 Arc-discharge scheme. Two graphite electrodes are used to produce a DC electric arc-discharge in inert gas atmosphere.

Iijima produced the first MWCNTs by this method. He found that nanotubes were formed on the cathode, along with soot and fullerenes. Iijima and Ichihashi and Bethune and co-workers were the first to report on the production of SWCNTs. Both Iijima and Bethune found that SWCNTs could only form by adding a metal catalyst to the anode; specifically, Iijima used an Fe:C anode in a methane:argon environment, while Bethune utilized a Co–C anode with a He environment.

There are several variations one can make to tailor the arc discharge process. Currently, most growth is carried out in an Ar–He gas mixture. By tailoring the Ar–He gas ratio, the diameter of the SWCNTs formed can be controlled, with greater Ar yielding smaller diameters. The anode–cathode distance can be changed to vary the strength of the plasma formed in between. The overall gas pressure has been shown to affect the weight per cent yield of SWCNTs. Several metal catalyst compositions produce SWCNTs, but the current standard for SWCNT production is a Y–Ni mixture that has been shown to yield up to 90% SWCNT, with an average diameter of 1.2 to 1.4 nm. In general, the nanotubes produced by this method need extensive purification before use.

Chemical vapour deposition (CVD)

Both hot-wall as well as cold-wall plasma-enhanced CVDs (PECVDs) have been developed for the synthesis of CNTs.

Hot-wall CVD method This process includes catalyst-assisted decomposition of hydrocarbons, usually ethylene or acetylene, in a tube reactor at 550–750°C and the growth of carbon nanotubes over the catalyst when the system is cooled. Fe, Co and Ni nanoparticles can be used as catalysts. These particles catalyse the breakdown of the gaseous molecules into carbon, and a tube then begins to grow with a metal particle at the tip (Fig. 6.9).

Large-scale synthesis of aligned carbon nanotubes can be achieved by the CVD technique with iron as catalyst. A substrate containing iron nanoparticles embedded in mesoporous silica can be placed in the reaction chamber. A mixture of 9% acetylene in nitrogen can be introduced in the chamber at optimum flow rates. With these conditions, carbon nanotubes will form on the substrate containing the iron nanoparticles by deposition of carbon atoms obtained by decomposition of acetylene at 700°C. Catalytic synthesis can be easily scaled up in comparison to arc evaporation. With CVD, SWCNTs from 0.4 to 5 nm can be readily produced, and depending on the conditions, feedstock and catalyst, the yield can exceed 99% (weight per cent of final material) and the final product can be completely free of amorphous carbon.

Cold-wall CVD synthesis This includes techniques based on both hot wire and plasma-enhanced (RF/DC/microwave/ECR plasma) CVD techniques. To grow SWCNT by hot-wire CVD (HWCVD), hydrocarbons with dissolved Fe-containing molecules are passed over an extremely hot filament near the furnace entrance to facilitate a plasma-induced breakdown of hydrocarbons and nucleation of nanotubes. By initiating growth in the vapour phase, a substrate can be placed downstream in a cooler area of the furnace (~450°C) so that the nanotubes can deposit from the vapour phase onto the substrate. This is useful because by lowering the substrate temperature, the variety of substrates on which SWCNTs can be synthesized is widened.

Plasma-enhanced chemical vapour deposition (PECVD) systems have been used to produce both SWCNTs and MWCNTs. PECVD is a general term, encompassing several differing synthesis methods. In PECVD, a glow discharge is generated in a chamber when a high voltage is applied between two electrodes. Figure 6.10 shows a schematic diagram of a typical plasma CVD apparatus with a parallel plate electrode structure.

In PECVD, a uniform film is obtained by the supply of reaction gas from a plate held opposite to the substate which is placed on the grounded electrode. Catalytic metals, such as Fe, Ni and Co, are used on a Si, SiO_2 or glass substrate using thermal CVD or sputtering. Carbon nanotubes grow on the metal nanoparticles by glow discharge from a high-frequency

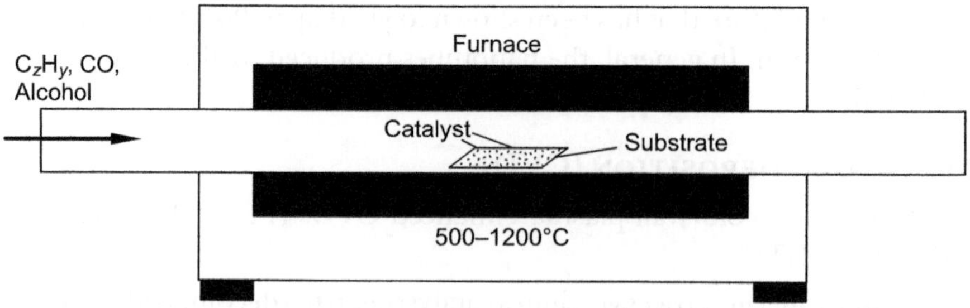

Fig. 6.9 Schematic of the hot-wall CVD reactor used for nanotube synthesis.

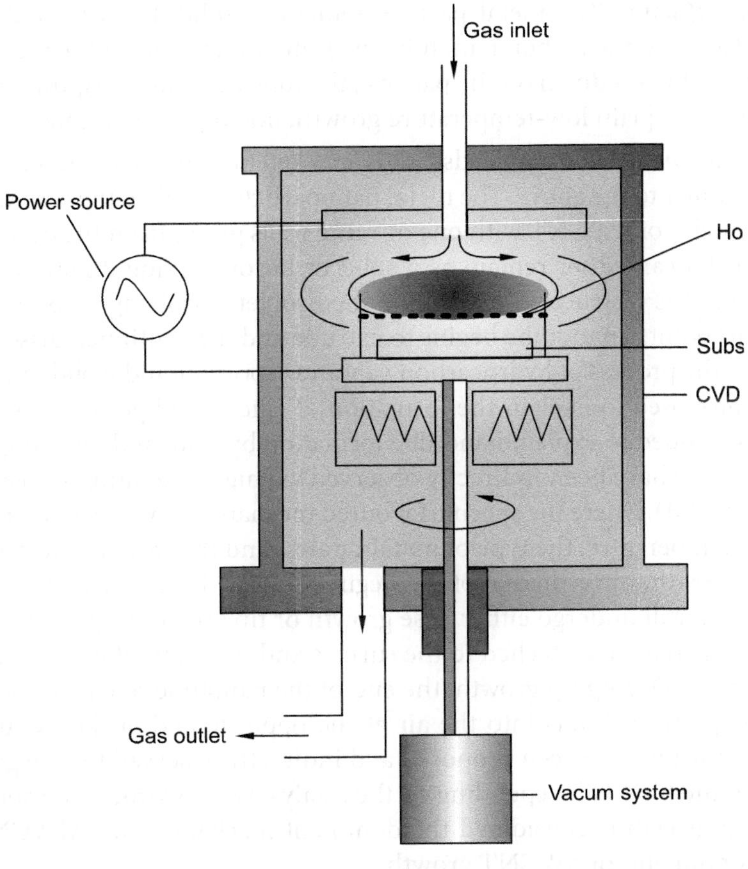

Fig. 6.10 Schematic diagram of a typical plasma CVD apparatus with a parallel plate electrode structure.

power. A carbon-containing reaction gas, such as C_2H_2, CH_4, C_2H_4, C_2H_6 or CO is supplied to the chamber during the discharge. The diameter of the nanotube, growth rate, wall thickness, and morphology are influenced by the catalyst. Ni seems to be the most suitable pure metal catalyst for the growth of aligned multi-walled carbon nanotubes (MWNTs). The MWNTs have a diameter of about 15 nm. The highest yield of carbon nanotubes achieved was about 50% and was obtained at relatively low temperatures (below 330°C). Direct PECVD systems can be used for the production of MWCNT field emitter towers and some SWCNTs.

CNT growth in CVD can be split into two basic types depending on the location of the catalyst: the so-called gas phase growth and substrate growth. Both these growth pathways can in turn be split into bulk carbon diffusion and surface carbon diffusion models. In gas phase growth, catalyst formation and nanotube growth occur literally in mid-air. In substrate growth, catalyst nanoparticles or metal precursors are deposited either on a substrate such as SiO_2 or on a high-surface area powder before growth. The underlying chemistry for both methods that leads to the formation of nanotubes from nanoparticles is similar, and both can usually be classified into surface carbon diffusion and bulk carbon diffusion.

Surface carbon diffusion The metal particle remains a solid, the 'cracked' carbon diffuses around the surface, and the carbon nanotube nucleates on the side of the metal particle. Since carbon continually breaks down on the particle, the tube continues to grow. This is a common mechanism used to explain low-temperature growth, notably with Ni catalyst nanoparticles.

Bulk carbon diffusion The carbon feedstock is 'cracked' or broken down on the surface of the metal particle, similar to the above. The metal nanoparticle dissolves the carbon until it reaches saturation, at which point a CNT with one or more walls grows from the outer surface. In this situation, the metal can either remain as a solid or become a liquid nano-droplet. Where it becomes a liquid, it is instructive to imagine the droplet dissolving carbon until it becomes saturated. At this point, a nanotube begins to extrude and the continued dissolution of carbon provides fuel for the process of hydrocarbon vapour (vapour–liquid–solid model). This model was originally proposed to explain the formation of silicon and germanium whiskers in the 1960s and was extended to explain nanotube formation by Saito and his group.

Both mechanisms have been indirectly observed via high-resolution transmission electron microscopy (HRTEM), where the specific favoured mechanism by a particular growth method depends on the temperature, the type of metal catalyst and the carbon feedstock used.

In substrate growth, once the nanotube begins to grow by either surface or bulk carbon diffusion, the CNT will undergo either base growth or tip growth (Fig. 6.11). In base growth, the catalyst particle remains attached to the surface and the nanotube is extruded into the air or along the surface. During tip growth, the end of the nanotube remains stuck to the surface and the catalyst particle shoots into the air at the opposite end of the extruding nanotube. These two mechanisms have been proposed and indirectly observed for the growth of carbon fibres, MWCNT and SWCNT, depending on the catalyst type, hydrocarbon source and growth temperature. Tip growth is considered the dominant mechanism for MWCNT growth, and base growth is dominant for SWCNT growth.

For a catalyst particle of unchanging size, the growth of CNTs should continue until the hydrocarbon is shut off, either by removing the feedstock from the reaction area or by

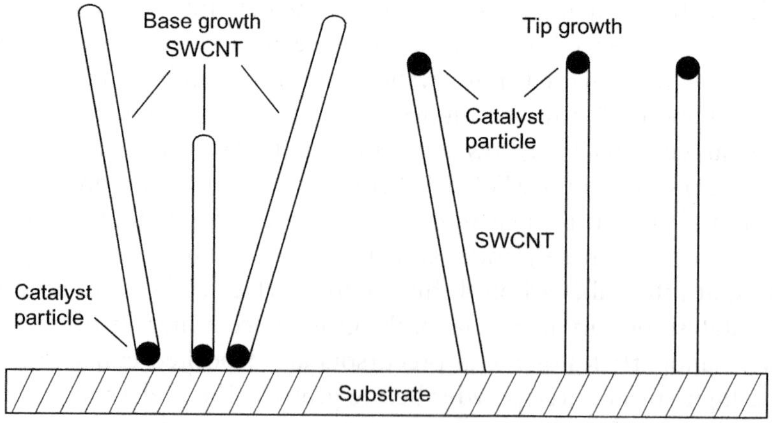

Fig. 6.11 Schematic showing the two types of growth mechanisms of a CNT.

amorphous or graphitic carbon fully coating the particle, blocking the gas. Additionally, in the case of base growth, growth may slow down or stop due to slow diffusion of hydrocarbons down to the nanoparticle at the bottom of the CNT. If nothing impedes the source of carbon to the nanoparticle, and nothing impedes the nanotube extrusion, growth should be continuous. In reality, there are competing reactions at the nanoparticle site, such as the formation of graphitic shells and the deposition of amorphous carbon. As a result, in sub-optimal growth conditions, amorphous carbon can coat the nanoparticle, preventing feedstock from reacting with the particle and cutting off the carbon source, thus terminating the growth. Alternatively, if the nanotube exitting the particle encounters an excessive external force, the energy required for forming a nanotube might exceed the energy necessary to form a graphitic carbon shell, at which point the nanoparticle will coat itself with a carbon shell, cutting itself off from the carbon feedstock.

High-pressure carbon monoxide method

One of the recent methods for producing SWCNTs in gram to kilogram quantities is the HiPCO process. The high yield demonstrates the high potential of this method for bulk production of SWNTs. Catalysts for SWNT growth form in situ by thermal decomposition of ion pentacarbonyl in a heated flow of carbon monoxide at pressures of 1–10 atm and temperatures of 800–1200°C. Previous methods for growing CNTs using hydrocarbons as the source have resulted in large quantities of amorphous carbon and graphitic deposits due to the thermal breakdown of hydrocarbons at high temperatures. The amorphous carbon overcoating would have to be removed in subsequent steps. The HiPCO method uses carbon monoxide as the carbon feedstock and $Fe(CO)_5$ as the iron-containing catalyst precursor. The SWNT yield and the diameter of the nanotubes produced can vary over a wide range determined by the condition and flow-cell geometry. The HiPCO method produces SWNTs by flowing CO mixed with a small amount of $Fe(CO)_5$ through a heated reactor. The products of the $Fe(CO)_5$ thermal decomposition react to produce iron clusters in the gas phase. SWNTs nucleate on these clusters and grow. The solid carbon is formed through CO disproportionation, also known as the Boudouard reaction:

$$CO + CO \longrightarrow C(s) + CO_2$$

Iron nanoparticles act as catalysts for the above reaction. The flow tube has a thick quartz wall and is contained within the furnace. The rate at which the reactant gases are heated determines the amount and quality of the SWNTs produced. The CO and $Fe(CO)_5$ gases are maintained at a low temperature initially through a water-cooled injector. This low temperature is maintained so that rapid heating can occur inside the furnace. This apparatus of the HiPCO method yielded high quantities of SWNTs. Though related to CVD synthesis, HiPco deserves particular mention, since in recent years it has become a source of high-quality, narrow-diameter SWCNTs for distribution around the world. The metal catalyst is formed in situ when $Fe(CO)_5$ or $Ni(CO)_4$ is injected into the reactor along with a stream of carbon monoxide (CO) gas at 900 to 1100°C and at a pressure of 30 to 50 atm. The reaction to make SWCNTs is the disproportionation of CO by nanometre-size metal catalyst particles. Yields of SWCNT material are claimed to have up to 97% atomic purity. The SWCNTs made by this process have diameters ranging between 0.7 and 1.1 nm. By tuning the pressure in the reactor and the catalyst composition, it is possible to tune the diameter range of the nanotubes produced.

The production of SWNTs by using CO as the sourse of C and $Fe(CO)_5$ as the iron-containing catalyst precursor is usually termed as the high-pressure CO disproportionation process (HiPCO). A gas mixture of CO and $Fe(CO)_5$ is passed through a heated reactor to produce SWNTs. Nanotubes as small as 0.7 nm in diameter, which are expected to be the smallest achievable chemically stable SWNTs, have been produced by this method. The average diameter of HiPCO SWNTs is approximately 1.1 nm. About 70% yield could be achieved by this process. Use of the highest possible temperature and pressure can lead to the highest yield and narrowest tubes. SWNT material with 97% purity can be produced at rates of up to 450 mg/h with this process.

Laser ablation

The laser ablation technique uses a 1.2 at.% of cobalt/nickel with 98.8 at.% of graphite composite target that is placed in a 1200°C quartz tube furnace with an inert atmosphere of ~500 torr of Ar or He and vaporized with a laser pulse. A pulsed- or continuous-wave laser can be used. Nanometre-size metal catalyst particles are formed in the plume of vaporised graphite. The metal particles catalyse the growth of SWCNTs in the plasma plume, but many by-products are formed at the same time. The nanotubes and by-products are collected via condensation on a cold finger downstream from the target. The yield varies from 20% to 80% of SWCNTs by weight. The by-products of this synthesis are graphitic and amorphous carbon, 'bucky onions' (concentric fulleriod spheres) surrounding metal catalyst particles and small fullerenes (C_{60}, C_{70}, etc.). In principle, arc discharge and laser ablation are similar methods, as both use a metal-impregnated graphite target (or anode) to produce SWCNTs, and both produce MWCNT and fullerenes when pure graphite is used instead. The diameter distribution of SWCNTs made by this method is roughly between 1.0 and 1.6 nm. This method has a yield of around 70% and produces primarily single-walled carbon nanotubes with a controllable diameter determined by the reaction temperature. However, it is more expensive than either arc evaporation or chemical vapour deposition.

Laser ablation gives high yields of SWNTs. The process involves laser ablation of graphite rods with small amounts of Ni and Co at 1200°C (Fig. 6.12). The synthesized nanotubes are remarkably uniform in diameter and they form ropes 5–20 nm in diameter and tens to hundreds of micrometres long. The growth of nanotubes takes place by *scooter mechanism* in which, a single Ni or Co atom chemisorbs on the nanotube. The metal atom must have sufficiently high electronegativity to prevent the formation of fullerenes and it must be highly effective in catalysing the nanotube growth. The metal atom circulates around the open end of the tube and absorbs small carbon molecules and converts them into graphite-like sheets. The tube growth terminates when many catalyst atoms accumulate at the end of the nanotube, poisoning the catalysis process.

In this process, a mixture of graphite and (Co, Ni) catalyst is vaporized by a laser beam in a tube furnace at 1200°C. The nanotubes are deposited on a collector (water cooled) kept outside

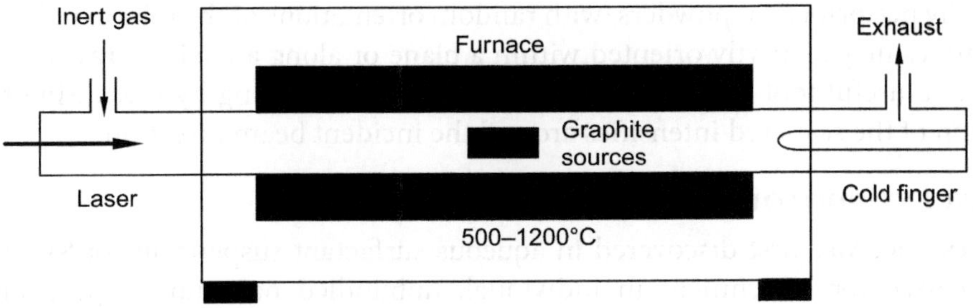

Fig. 6.12 Schematic drawing of a laser ablation apparatus.

the furnace. The laser ablation yields >70% highly endless, highly tangled ropes of SWNTs along with nanoscale impurities. The potential use of the nanotubes in nano-electronic devices requires nanotubes with length in the range of 10–300 nm.

Purification of nanotubes

The different methods of production of nanotubes suffer some serious limitations; all produce mixtures of nanotubes and nanoparticles sticking together in larger lumps. In addition, the length of the tubes varies widely and the nanotubes have a number of defects in them. Purification is basically the separation of tubes from the soot and elimination of defects by different post-growth treatments. Ultrasonication is one of the techniques by which the tubes are freed from particles. While the larger contaminants can be removed by sedimentation in a solvent, it is not easy to remove the smaller particles. One possibility for MWNTs is to perform an oxidative treatment, either by heating the powder in air at 650°C or by a liquid phase treatment in an acidic environment.

Purification of SWNTs can be done by fluxing in acid followed by centrifuging. Microfiltration or size-exclusion chromatography, which does not use acids, is also used for SWNTs. In case of MWNTs, a colloidal suspension is prepared in which the larger particles sediment, while the smaller remain in suspension. Size-exclusion chromatography was also used successfully for MWNTs. Heating the carbon nanotubes up to 3000 K can also eliminate impurities and defects in the tubes. Separating nanotubes with different chirality is more complex. Conducting nanotubes can be separated from semiconducting ones by the application of current between metal electrodes, which causes conducting nanotubes to burn, leaving the semiconducting ones.

6.2.4 Characterization techniques in carbon nanotubes

Characterization tools are crucial in the study of emerging materials to evaluate their full potential in applications and to comprehend their basic physical and chemical properties. Typically, a wide range of tools are brought to bear in order to elicit these properties.

X-ray diffraction

XRD measurements are routinely employed to characterize carbon nanotubes. Nanotubes show only (hk0) and (00l) peaks in XRD patterns. Neutron and X-ray diffraction experiments are generally performed on powders, with random orientations of the tube axis. In other cases, the nanotubes may be partly oriented within a plane or along an axis. X-ray diffraction may then reveal a useful tool to characterize their degree of ordering by measuring the angular distribution of the scattered intensities around the incident beam direction.

Fluorescent photoluminescence

The fluorescence was first discovered in aqueous surfactant suspensions of SWNT that had been processed for enrichment in individual, unbundled nanotubes. Spectrofluorimetric measurements of emission intensity as a function of excitation and emission wavelengths revealed a rich pattern of peaks representing distinct (n, m) structural species. This spectral assignment provided a large body of precise optical transition energies for a significant range of tube diameters and chiralities. Important patterns of electronic structure emerged, showing the related properties of nanotubes within 'families' (sharing the same n–m value) and 'tribes' [sharing the same mod (n– m, 3) value]. Spectral line shapes reveal the predominant excitonic character of optical excitations in SWNT. Nanotube fluorescence is quenched by aggregation, chemical derivatisation, and by acidification in some aqueous suspensions. Fluorimetry offers a powerful method for determining the (n, m) composition of mixed nanotube samples. Fluorescence-based spectroscopic studies provide precise information about electronic excitation energies and their dependence on nanotube structure, while line shape analyses and time-resolved emission experiments give insights into relaxation processes. The spectrofluorimetric ability to monitor in parallel a wide range of (n, m) species allows studies of processes in which nanotubes are chemically or physically manipulated in structure-dependent ways.

Electron microscopy

Using different electron microscopy techniques, it is possible to study carbon nanotube structures in great detail and identify their growth mechanism, which in turn helps to gain an insight into improving the nanotube growth processes or modifying their structure.

Scanning electron microscopy (SEM) SEM allows imaging of the ropes of single-walled nanotubes (SWNTs) in a sample or to view the highly oriented forest of multi-walled carbon nanotube films grown on quartz or silicon substrates (Fig. 6.13). The typical resolution of an SEM is about 1 to 5 nm. Although the resolution of this technique does not allow for individual SWNTs to be imaged within a SWNT bundle, combined with other techniques, it can be used to determine the amount of impurities such as amorphous carbon or carbon-coated catalyst particles, which typically co-exist with SWNT bundles in the sample. SEM can be useful in imaging the tubular 1D structure of the MWNTs.

Transmission electron microscopy (TEM) This is a powerful technique that allows one to determine the number of walls in an MWNT or image the isolated SWNTs residing inside an SWNT bundle. This allows for careful measurement of tube diameters as well as investigation of structural defects in carbon nanotubes. Many studies have been done on nanotubes involving TEM. TEM observations clearly reveal that the carbon nanotubes prepared by the arcing

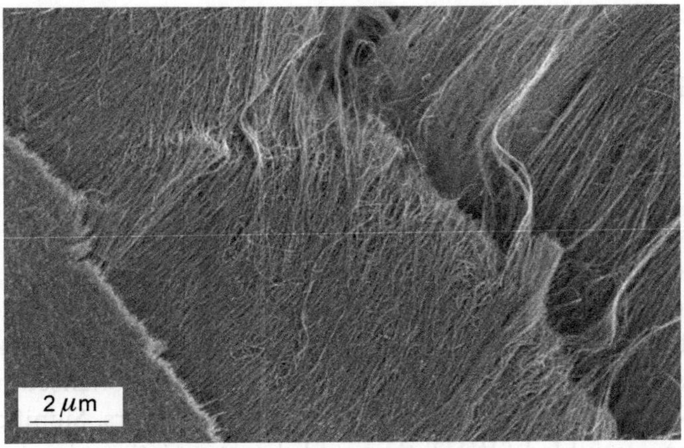

Fig. 6.13 Multi-walled carbon nanotube (MWCNT) sample made by a CVD process using iron-containing catalytic particles. The MWCNTs adhere in mats. (*Source:* http://commons.wikimedia.org/wiki/File:Fe_cat_CVD_MWCNT_SEM_overview_mats_with_scalebar.jpg).

process generally consist of multi-layered, concentric cylinders of single graphitic (graphene) sheets. The diameter of the inner tubes is generally of the order of a few nanometres. The outer diameter could be as large as 10–30 nm. Nanotubes get their helicity during the curling of the graphene sheet into a cylinder. Electron diffraction patterns show the presence of helicity, suggesting that the growth of nanotubes occurs as in the spiral growth of crystals. The concentric cylinders in MWNTs are about 0.345 nm apart, which is close to the separation between the (002) planes of graphite, which are the lowest energy surfaces of graphite with no dangling bonds. In the electron microscope images, one typically observes nanotubes along their lengths, with the electron beam falling perpendicular to the axis of the nanotube. Iijima has published such an image for the (110) planes separated by 0.21 nm. Ring-like patterns are found due to individual tubes that consist of cylindrical graphitic sheets that are independently oriented (with no registry between the sheets) with helical symmetry for the arrangement of the hexagons.

Atomic force microscope (AFM)

AFM has been very useful in imaging isolated SWNTs that have been grown directly on silicon substrates via a chemical vapour deposition process. Interestingly, nanotubes have been used as AFM tips. Their 1D nature allows them to be used as an improvement over existing AFM tips. Also, nanotubes tend to buckle under mechanical pressure, which will allow a nanotube AFM tip to survive a crash into a sample with little or no damage to the tip. Thus, not only are nanotubes imaged using AFM, but they also possess qualities that can improve the technique as well.

Electron beam microscopy and scanning tunnelling microscopy (STM) require a carefully controlled vacuum environment and rely on specific sample limitations. Because the AFM relies on a mechanical response for its data acquisition, it is uniquely suited to work in ambient conditions and is not limited to conductive samples. AFM's ability to achieve atomic-scale resolution is based on a number of refinements:

- Flexible cantilevers
- Sensitive detection
- High-resolution tip–sample positioning
- Tip sharpness
- Force feedback

The smaller diameter, high aspect ratio and large Young's modulus make carbon nanotubes suitable materials for AFM tips. Nanotube probes in AFM have been used to observe a number of new biological structures. The 'chemical resolution' of functionalised manually assembled MWNT and SWNT tips has been tested on partial lipid bilayers. Significantly, these studies have shown that an assembled SWNT tip could detect variations in chemical functionality with resolution down to 3 nm, which is the same as the best structural resolution obtained with this type of tip. This resolution should improve with CVD SWNT tips.

Raman spectroscopy of nanotubes

Raman spectroscopy is a powerful, multifaceted technique with wide-ranging applications in carbon nanotube studies. It is also capable of identifying nanotube electronic nature through analysis of the nanotube G-band, found near 1600 cm^{-1}. Semiconducting nanotubes produce a two-peak G-band structure (G+ and G– bands) with Lorentzian line shapes, while the G– peak in metallic types shows a pronounced broadening with a Breit–Wigner–Fano (BWF) line shape. (This line shape is caused when the tangential phonon couples to the continuum of electronic states at the Fermi level.) Additionally, measurement of the nanotube D-band (found near 1300 cm^{-1}) can provide an evaluation of the defect density occurring in a sample and act as a useful tool for monitoring the extent of covalent sidewall modification in nanotube functionalisation chemistry.

Raman spectroscopy becomes a sensitive probe of nanotube electronic structure and coupling of phonons to electronic transitions. Strong resonance enhancement is obtained from nanotubes through tuning of the excitation wavelength to overlap the van Hove singularities present in the 1D density of states of different nanotube species. Because the electronic resonances are dependent on nanotube diameter and chirality, only a subset of the total nanotube population is accessed at any given excitation wavelength.

6.2.5 CNT-based physical sensors

Nanotube-based physical sensors can measure pressure, flow, temperature or the mass of an attached particle. Researchers at the Georgia Institute of Technology have demonstrated a carbon nanotube–based nanobalance which can weigh submicron scale particles. By applying an alternative voltage, they were able to create resonance in the nanotube with a specific frequency, which depends on the nanotube length, diameter, density and elastic properties. The mass of the particle was calculated from changes in the resonance frequency that occurs on placing the particle. Based on this technique, researchers were able to measure the mass of a carbon sphere as 22 femtograms, which is by far the smallest mass ever measured. This approach may lead to a technique for the measurement of the weight of individual biomolecules. Beyond the small particle measured so far, researchers believe their nanobalance could be useful for determining the mass of other objects in the femtogram to picogram range; for example, viruses.

Cleland and Roukes at the California Institute of Technology have reported the fabrication and characterisation of a working *nanometre-scale mechanical electrometer*. The device has demonstrated charge sensitivity below a single electron charge per unit bandwidth (~0.1 electrons/Hz at 2.61 MHz), which is better than that of state-of-the-art semiconductor devices and comparable with the charge detection capabilities of cryogenic single electron transistors.

Single-walled carbon nanotubes have been shown to exhibit *piezoresistive effect*, i.e., when they are bent or stretched, their electrical resistance will change. Based on this principle, carbon nanotube–based pressure and strain sensors have been developed. The pressure sensor consists of an ultrathin aluminium oxide membrane to which carbon nanotubes are attached. Christoph Stampfer of ETH in Zurich and his co-workers put these principles into practice to make a well-characterized nanotube-based pressure sensor. They also used the idea of placing a nanotube, connected at each end to electrodes, on an ultrathin membrane, so that when the membrane bends or bulges, the nanotube bends too. To calibrate the device, they measured the deformation of the membrane in response to an applied pressure by using white-light interferometry. They then monitored changes in nanotube resistance as a function of strain. They could detect a change even for strains as small as a hundredth of a per cent, which in this case were induced by pressures of a few tens of kilopascals. The sensing nanotube was in this case metallic so that the gauge factor was positive. It had a value close to that of the best silicon devices.

SWNT-based flow sensors produce electrical signals in response to fluid flow directly. This is due to the direct scattering of the free carriers from the fluctuating Coulombic fields of the ions or polar molecules in the flowing liquid. It was found that the ionic strength of the flowing liquid significantly affected the induced voltage. The electronic properties—such as the local density of states—of single-walled carbon nanotubes have been shown to be extremely sensitive to the chemical environment, because in this case all the tube atoms are surface atoms. The electrical conductivity and thermoelectric power also vary during exposure to several gas species and most of the nanotube-based chemical sensors have been developed based on this principle.

Kong and his co-workers showed that the electrical resistance of individual semiconducting SWNTs changes by up to three orders of magnitude on exposure to NO_2 or NH_3 molecules at room temperature. Based on this they devised a chemical sensor in which a single semiconducting SWNT was placed in contact with titanium or gold metal electrodes at the ends. SWNT-based hydrogen sensors have also been demonstrated by sputter-coated individual and bundled SWNTs with Pd nanoparticles. The conductivity was found to decrease by 50% and 33% for individual and nanotube bundles on exposure to air mixed with

400 ppm of hydrogen. Oxygen sensors based on change in resistance and thermoelectric power on exposure to oxygen have been reported.

For biosensors, various ligands can be connected to a nanotube transducer in order to recognise specific target molecules. The large surface areas available for molecule adsorption make carbon nanotubes a suitable material for biosensors. Carbon nanotube–based sensors are a thousand times smaller than microelectromechanical system (MEMS) sensors and consume less power. These two important advantages will make them useful as implantable devices. Such biosensors based on functionalized carbon nanotubes will provide high sensitivity, large linear range, fast response, long life and low detection thresholds for different analytes.

6.3 GaN NANOWIRES

GaN is an important wide-band gap semiconductor with a variety of applications in electronic and opto-electronic devices. Single crystal GaN nanowires with the wurtzite structure have been prepared by several methods, the simplest being the thermal evaporation of GaN powders at a temperature of 1200°C in an Ar atmosphere without the use of a template or a patterned catalyst. The as-synthesized nanowires have a diameter of 30 nm and length of several microns. Dielectric property measurements on GaN nanowires at room temperature at different frequencies reveal that the dielectric constants of the nanowires are much larger than those of powders at low frequencies, which can be explained in terms of space charge polarization and rotation direction polarization. The sublimation method has also yielded bamboo-shaped structures in an Ar–NH_3 atmosphere at 930°C, starting from GaN powder. Sublimation of GaN powder yielded GaN nanowires of 10–45 nm diameter on $LaAlO_3$ substrates. Hot filament chemical vapour deposition, wherein a solid source of Ga_2O_3 and C mixture was heated in a NH_3 atmosphere, has been employed to synthesise bulk quantities of GaN nanowires of 5–12 nm diameter and several micrometres in length. PL spectra of the nanowires show a broadband around 420 nm, possibly arising from the blue shift of the yellow luminescence in bulk GaN.

Synthesis method with experimental parameters	Resulting wire or tube features
Laser ablation: GaN/Fe target, 250 torr, 900°C	~10 nm diameter, >1 μm long, [1 0 0] growth direction
HFCVD: Ga_2O_3/C powder, NH_3, 200 torr, 900°C,1 hour	5–12 nm diameter, >1 μm long, single crystalline.
Direct reaction of Ga with NH_3: 825–925°C, 15 torr, 3–4 hours	20–150 nm diameter, 500 μm long

Hydride vapour phase epitaxy has been carried out employing a sapphire substrate involving $GaCl_3$ mixed with flowing NH_3 at a temperature of 478°C to obtain **nanorods**. Cathodoluminescence (CL) measurements of individual GaN nanorods show a blueshift with decrease in the diameter of GaN nanorods due to quantum confinement. GaN nanowires of 10–50 nm diameter, with a minimum diameter of 6 nm, have been synthesized by the use of molten Ga and various catalysts, in the temperature range of 850–1000°C over Si wafers and quartz, by the VLS mechanism. Temperature-dependent PL measurements show a strong band-edge emission and a weak yellow luminescence at 2.3 eV. Field emission studies on nanowires have shown significant emission currents at low electric field with current densities of 20 $\mu A/cm^2$ at a field of 14 V/μm.

The synthesis of straight and smooth GaN nanowires of 10–40 nm diameter and 500 lm length starting from Ga and NH_3 at temperatures of 920–940°C has been achieved, by the use of catalytic nickel oxide nanoparticles dispersed over $LaAlO_3$ substrates. Laser-assisted catalytic growth of a GaN/Fe target has been employed successfully to obtain nanowires of 10 nm diameter and 1 lm length, wherein the in situ generation of the fine liquid nanoclusters of the catalyst causes the growth of the nanowires. The growth direction of the GaN nanowires is [100], running parallel to the wire axis. HFCVD of a mixture of Ga_2O_3 and C powders under flowing NH_3 at around 900°C yields GaN nanowires in large quantities. The growth of the nanowires can be explained by the VS mechanism, where the nanowires tend to form on different planes depending on the temperature since the reaction is far from thermodynamic equilibrium conditions. Laser ablation of a composite target of Ga_2O_3 and GaN powders gives core-sheath GaN nanowires of hexagonal and cubic structure. The growth of the nanowires is attributed to the oxide-assisted mechanism consisting of a series of oxidation and reduction reactions occurring during the nucleation and growth process.

Large-scale synthesis of GaN nanorods has been carried out by the direct reaction of metallic gallium vapour with flowing ammonia at 970°C. PL measurements show two strong peaks at 377 nm (3.28 eV) and 360 nm (3.44 eV) attributed to the zero phonon donor–acceptor pair transition and the donor-bound exciton, respectively. GaN nanorods of 4–50 nm diameter and lengths extending to 25 lm were obtained through a carbon nanotube confined reaction where Ga_2O vapour was reacted with NH_3 gas. GaN nanowires are also obtained by heating $Ga(acac)_3$ in the presence of carbon nanotubes or activated carbon in NH_3 vapour at 910°C. The reaction can also be carried out by heating $Ga(acac)_3$ in NH_3 vapour over catalytic Fe/Ni particles dispersed on silica. In the above reactions, the $Ga(acac)_3$ precursor generates fine Ga_2O particles in situ which then reacts with the NH_3 vapour. The diameter of the nanowires could be reduced to 20 nm by employing single-walled nanotubes (SWNTs) or a lower proportion of the catalyst.

CVD synthesis of high-quality ultrafine GaN nanowires has been accomplished by the use of hot-filament chemical vapour deposition, where liquid Ga placed over a p-type Si wafer and covered with a Ni film was heated at 700°C in NH_3. The growth of the nanowires is attributed to the VLS mechanism. CVD has also been employed with a Si/SiO_2 substrate coated with 30 Å Au via e-beam evaporation. GaN nanowires with diameter of 20–100 nm were grown on this substrate by using a Ga source heated to 900°C in flowing NH_3. Such a direct and precise control of the diameter of the nanowires is desirable in order to exploit the electronic and opto-electronic properties, which are closely related to the diameter. The method has been modified by employing e-beam lithography to create catalyst islands in specific patterned regions to enable the growth of the single GaN nanowire between two catalyst islands. This position-controlled growth is important for making nanowire integrated systems since one does not need to locate the nanowires individually. Many devices can be made in parallel on one chip by exploiting the advantage of batch processing. Thermal CVD has been employed to synthesise nanowires of diameter 50–60 nm on a NiO catalysed alumina substrate starting from Ga and GaN powders at a temperature of 1000°C in NH_3. Raman spectra of GaN nanowires reveal significant broadening of the Raman modes, which is indicative of the phonon confinement effects associated with the nanoscale dimension of the system.

A pyrolysis route has been employed for the synthesis of nanorods of GaN, wherein the pyrolysis of gallium dimethylamide and ferrocene was carried out in NH_3 using a two-stage furnace. The controlled growth of oriented GaN nanopillars and randomly distributed nanowires has been accomplished by MOCVD using $(N_3)_2Ga[(CH_2)_3NMe_2]$ in the presence of pure N_2 gas at high flow rates of 100 sccm at 950°C. The room temperature PL spectrum of the suspension of the GaN nanopillars showed a strong broad emission peak around 430 nm.

Large-scale preparation of GaN nanowires of diameter ~50 nm is possible starting from Ga and Ga_2O_3 using catalytic indium nanoparticles, within the nano-channels of anodic alumina templates. Doped GaN nanowires are of great interest due to their interesting optical and magnetic properties. Both p-type (Mn- and Mg-doped GaN) and n-type (Si-doped GaN) have been investigated. Ferromagnetic Mn-doped GaN nanowires have been prepared by reacting a mixture of acetyl acetonates with NH_3 at 950°C in the presence of multi-walled and single-walled carbon nanotubes, the nanowires prepared with SWNTs being considerably smaller in diameter (25 nm). GaMnN nanowires with 1%, 3% and 5% Mn have been shown to be ferromagnetic even at 573°C, exhibiting magnetic hysteresis. The GaMnN nanowires show higher coercivity than thin films.

6.3.1 Applications of GaN nanowires

Ultraviolet-blue laser action has been reported in monocrystalline GaN nanowires, using both near field and far field optical microscopy to characterise the waveguide mode structure and the spectral properties of the radiation at room temperature.

Field effect transistors (FETs) based on individual GaN nanowires have been fabricated. Gate-dependent electrical transport measurements show that the GaN NWs are n-type and the conductance of NW-FETs can be modulated by more than three orders of magnitude. Electron mobilities are comparable to or larger than thin film materials with similar carrier concentration and thus demonstrate the high quality of these nanowire building blocks and their potential for nanoscale electronics.

6.4 NANOCRYSTALLINE ZnO

ZnO is not a newly discovered material but is currently emerging as a promising candidate for semiconducting, piezoelectric and optoelectronic devices, owing to its optical and electrical properties, i.e., wide band gap (3.3 eV) and large excitation binding energy (60 meV).

NANOWIRE FABRICATION

Nanowire fabrication can be done by the following methods:

- Template assistance
- Electrochemical deposition
  - Ensures fabrication of electrically continuous wires
  - Occurs on conductive surfaces
  - Applicable to a variety of materials
- High-pressure injection
  - Limited to elements and heterogeneously melting compounds with low melting point
  - Cannot give continuous wires
  - Does not work well if the diameter is < 30–40 nm
- Chemical vapour deposition
- Laser-assisted techniques

Template-assisted synthesis

The template synthesis of nanowires is a conceptually simple and intuitive way to fabricate nanostructures. These templates contain very small cylindrical pores or voids within the host material, and the empty spaces are filled with the chosen material, which adopts the pore morphology, to form nanowires. The chemical stability and mechanical properties of the template are crucial in template-assisted synthesis of nanostructures. Templates frequently used for nanowire synthesis include anodic alumina (Al_2O_3), nano-channel glass, ion track-edged polymers and mica films. The template-assisted growth method can be carried out by:

- Pressure injection
- Electrochemical deposition
- Vapor deposition

VLS method for nanowire synthesis

VLS mechanism involves the absorption of the source material from the gas phase into a liquid droplet of catalyst. Upon supersaturation of the liquid alloy, a nucleation event generates a solid precipitate of the source material. This seed serves as a preferred site for further deposition of material at the interface of the liquid droplet, promoting the elongation of the seed into a nanowire, and suppressing further nucleation events of the same catalyst. Since the liquid droplet catalyses the incorporation of material from the gas source to the growing crystal, the deposit grows anisotropically as a whisker the diameter of which is dictated by the diameter of the liquid alloy droplet.

The nanowires thus obtained are of high purity, except for the end containing the solidified catalyst as an alloy particle.

Reduction of the average wire diameter to the nanometre scale requires the generation of nano-sized catalyst droplets. However, due to the balance between the liquid–vapour surface free energy and the free energy of condensation,,the size of the droplet in equilibrium with its vapour is usually limited to the micrometre range. These obstacles can be overcome by several new methodologies. Advances in the synthesis of metal nanoclusters have made monodispersed nanoparticles commercially available. These can be dispersed on a solid substrate in high dilution so that when the temperature is raised above the melting point, the liquid clusters do not aggregate. Metal islands of nanoscale size can self-form when a strained thin layer is grown or heat-treated on a non-epitaxial substrate. Laser-assisted catalytic VLS growth is a method used to generate nanowires under non-equilibrium conditions. By laser ablation of a target containing both the catalyst and the source materials, plasma is generated, from which catalyst non-clusters nucleate as the plasma cools down. Single crystal nanowires grow as long as the particle remains liquid. By optimisation of the material properties of the catalyst–nanowire system, conditions can be achieved for which nanocrystals nucleate in a liquid catalyst pool supersaturated with the nanowire material, migrate to the surface due to a large surface tension, and continue growing as nanowires perpendicular to the liquid surface.

The stable wide band gap at high temperatures and harsh chemical environment allows electronic and opto-electronic devices of ZnO to operate at high temperatures and hostile environments. The large excitation binding energy which is nearly 2.4 times the room temperature (RT) thermal energy makes ZnO an efficient excitonic emitter at room temperature as well as at high temperatures. The stability of ZnO to high-energy radiation, amenability to wet chemical etching and easy etching by all acids and alkalis has made it preferable to other wide band gap materials for the miniaturization of electronic, optical devices and further development of nano-devices. The existence of ZnO with the same crystal structure and close lattice parameter as GaN makes it a possible substrate for epitaxial film growth of the latter. ZnO has potential advantages over GaN for various applications, including its commercial availability, and also as a brighter light emitter than GaN. ZnO is also finding application in other areas like transistors. In the fabrication of transparent thin film transistors, the transistor is usually protected from light exposure by a covering. This problem is overcome in ZnO-based transistors as they are insensitive to visible light. ZnO is useful for making transparent electrodes in flat-panel displays and solar cells due to the possibility of changing electrical properties while maintaining optical transparency with controlled doping. Apart from this, it is a promising candidate for spintronics applications with a predicted Curie temperature of more than 300 K for Mn-doped ZnO.

6.4.1 Crystal structure and properties of ZnO

Zinc oxide *exists over a range of crystal structures* from hexagonal (wurtzite, B4) to zinc blende (B3) and rocksalt (B1) . Among all structures, wurtzite is thermodynamically stable at ambient conditions. At pressures around 9 GPa, ZnO exists in cubic rocksalt whereas the zinc blende structure stabilises when it has been grown on cubic substrates. At sufficiently high pressures (~ 260 GPa), one can observe structural changes in ZnO, from six-fold cubic rocksalt (B1) to eight-fold cubic CsCl (B2) structure. ZnO wurtzite structure has a hexagonal unit cell with two lattice parameters (*a*, *c*) and a *c*/*a* ratio of 1.633 which belongs to the space group of $P6_3mC$. The structure is composed by the interaction of two hexagonal close-packed sub-lattices; each unit cell consists of four atoms of one kind (Group II) surrounded by four atoms of another kind (Group VI) and vice versa, where all these coordinate at the edges of the tetrahedron. However, in reality, the Wurtzite structure deviates from the idea by changing the *c*/*a* ratio.

The Coulombic interaction caused by pressure-induced reduction in lattice parameter results in change in crystal structure of ZnO, from wurtzite structure to cubic rocksalt, which is quite normal in II-VI semiconductors. However, the six-fold coordinated rocksalt structure is metastable over long periods of time even at zero pressure. The reversal from rocksalt to wurtzite is also possible at around 1.5 to 2 GPa while decreasing the pressure from 10 GPa.

Zinc blende ZnO structure is also *metastable* and can be achieved by hetero-epitaxial growth on cubic substrates like ZnS, GaAs/ZnS and Pt/Ti/SiO_2/Si. Both zinc blende and wurtzite structures can exist together when ZnO is grown on highly mismatched substrates but separated by the crystallographic defects. Zinc blende structure is formed by the interpenetration of two face-centred sub-lattices shifted along the body diagonal by a length one quarter the body diagonal. Though wurtzite and zinc blende have the same tetrahedral coordination, the main difference between them is the stacking sequence: wurtzite consists of alternate close-packed

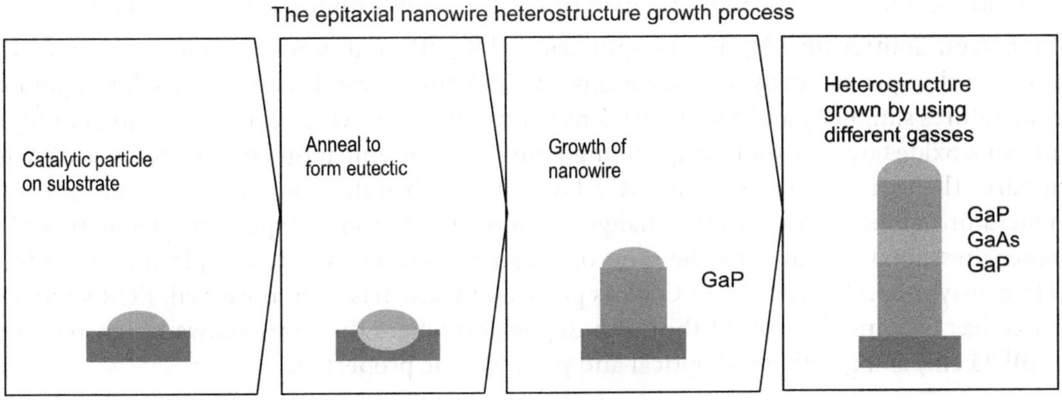

Fig. 6.14 Growth of III-V group nanowires by low pressure metal organic vapour phase epitaxy (MOVPE) on suitable crystalline substrates, such as III-V materials or silicon with a reasonably matching lattice constant. (*Source:* http://commons.wikimedia.org/wiki/File:MOVPEofEPINW.jpg)

(0001) planes arranged triangularly, whereas zinc blende consists of close-packed (111) planes arranged triangularly.

6.4.2 Synthesis of bulk-structured and nanostructured ZnO

Bulk single crystal growth

High-quality ZnO single crystals are increasingly required in optical devices. Due to its high vapour pressure, it is very difficult to grow ZnO from the melt, and the control over deposition is difficult in vapour phase deposition. The lower supersaturation favours the hydrothermal reaction for ZnO single crystal growth. In hydrothermal growth, ZnO seeds and sintered ZnO strings are heated together with an aqueous solution of KOH and LiOH at 300–400°C and pressures 70–100 MPa in a platinum crucible placed inside a two-zone vertical furnace. The inner Pt helps in reducing the contamination of ZnO with the aqueous solution.

Vapour phase transport gives high-quality ZnO wafers. This involves vaporization of ZnO power in a horizontal tube maintained at two different temperatures at the ends. Zinc oxide powder at 1150°C is transported as O and Zn with H_2 as the carrier gas, because of lower vapour pressure compared to ZnO. Once the Zn and O reach the other end maintained at 1100°C, the reverse reaction takes place with the help of single crystal seed, and transparent single crystal ZnO grows. Melt growth is the other method for producing bulk single crystal ZnO (Fig. 6.15). Melt growth mainly contains an induction heating set up with an RF source to melt the zinc oxide powder placed in a crucible with water cooling. Once all the powder melts, the crucible is taken from the heating set up and the melt is allowed to crystallise.

Thin Films ZnO thin films are being developed for transparent electrodes in new generation opto-electronic devices. Various deposition techniques like pulsed laser deposition, RF magnetron sputtering, chemical vapour deposition and molecular beam epitaxy are used to deposit nanocrystalline ZnO thin films. The films generally have a tendency to grow with (0001) as the preferred orientation on various substrates like glass, sapphire and diamond. Optically pumped lasing at room temperature is observed with high-quality ZnO films.

Magnetron sputtering Magnetron sputtering (DC, RF and reactive sputtering) is a well-known technique for early investigations of ZnO thin films because of its low operating temperatures simplicity and lower cost. ZnO films can be grown directly sputtering out highly pure zinc oxide target or sputtering out high pure metallic zinc target in the presence of Ar/O_2 mixture. The variant parameters are RF/ DC power, substrate temperature and gas pressure. In literature, authors observed the change in conductivity from n-type to moderate p-type (at higher pressures) by adjusting the ratio of O_2/Ar ratio in the sputtering plasma. In addition to this, p-type conductivity in ZnO films grown on GaAs has been observed. ECR sputtering source has been used for ZnO thin film deposition where the films showed high resistivity ($> 10^{10}\Omega$ cm) along with good optical and piezoelectric properties.

Molecular beam epitaxy (MBE)

It is used to obtain high-quality single crystal ZnO thin films by evaporating highly pure Zn metal from an effusion cell in oxygen plasma generated by and RF/ECR source. RF sources

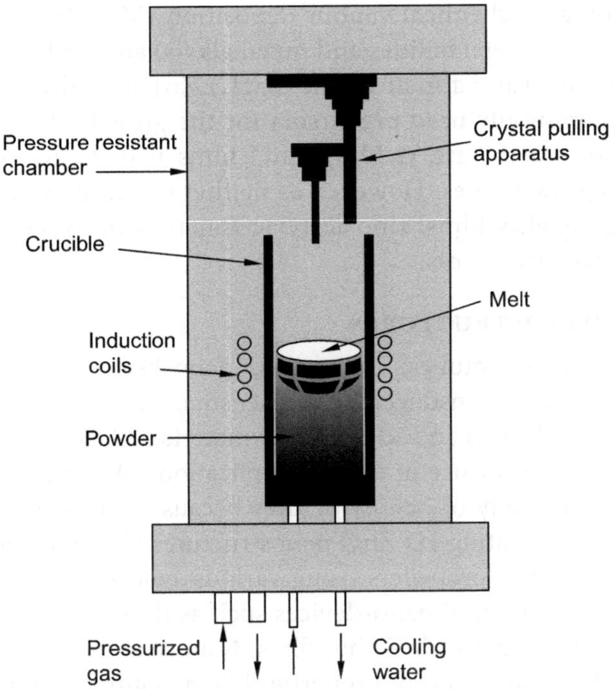

Fig. 6.15 Schematic diagram of the melt growth system.

are most commonly used because of the high oxygen radical density. Nitrogen dioxide and hydrogen peroxide are also used in MBE to oxidize Zn films. The main advantage of MBE is good control on the deposition conditions.

Pulsed laser deposition (PLD)

This is one of the promising techniques to grow nanocrystalline ZnO thin films by ablating a highly pure ZnO target in the presence of O_2 gas. Excimer lasers and Nd:YAG pulsed lasers are the most commonly used for ZnO target ablation. The advantages of PLD over other techniques are creation of high-energy source particles, preserving stoichiometry of the material and high-quality film growth at low substrate temperatures. Laser intensity, ambient oxygen pressure and substrate temperature are the operating parameters for deposition of ZnO thin films by PLD. In literature, authors developed a two-stage deposition process which involves nucleation of ZnO at lower pressures (10^{-4} torr), which acts as a high-quality template for single crystal growth at high pressures (10^{-1} torr).

Metal organic chemical vapour deposition (MOCVD)

The CVD deposition technique is gaining importance in the deposition of thin films, mainly because of its suitability in large-scale applications. Depending on the precursor gases and energy sources, they have been classified into various groups. The most frequently used technique is metal organic chemical vapour deposition (MOCVD). As the name implies, metal organic precursors (metal halides and metal alkyls) are used to deposit the thin films. $ZnCl_2/ZnI_2$ (metal halides) and dimethyl zinc $[(CH_3)_2Zn]$ or diethyl zinc $[(C_2H_5)_2Zn]$ metal alkyls are the most commonly used precursors for the growth of ZnO thin films. Among halides, iodide-based systems are yielding ZnO films with better optical and structural properties with high growth rates. However, as neither of the above mentioned metal alkyl precursors gives high quality films, zinc acetylacetonate with oxygen was recently used to obtain high-quality ZnO thin films.

Growth of ZnO nanostructures

Nanostructures such as nanotubes, nanowires, nanosheets and nanorods are gaining importance in fabricating transducers, nanosensors, nano-resonators, actuators, field emitters, opto-electronic devices and so on. Many materials like C, Si, InP, GaAs, SnO_2, ZnO, GaN and InO_3 are available for use in various applications. Among all these materials, ZnO probably has the richest family of nanostructures because of its electrical properties. Much work has been done in depositing 1D ZnO nanostructures like nanowires, nano-belts, nano-rings, nano-donuts and nano-propellers using various synthesizing routes. There are many reports claiming the fabrication of nano-devices such as thin solar cells with highly oriented ZnO nanostructures and textured ZnO thin films. Nanostructures form the base for various novel materials and the exploitation of properties like quantum confine occurs at nanoscale. Semiconductors based on nanostructures are finding application in diverse areas, from drug delivery to photovoltaic devices. Synthesis parameters like deposition temperatures, pressures and gas flux play a pivotal role in getting various ZnO nanostructures. Figure 6.16 shows various ZnO nanostructures such as hexagons, wires, tetrapods, spheres and brushes grown by different synthesis routes by Prof. Ramachandra Rao and his group at IIT Madras.

Vapour transport technique is most commonly used for the growth of ZnO nanostructures. This technique involves the transportation of Zn and O vapours and reaction with each other to give ZnO nanostructures. There are many ways of producing ZnO vapour. It can be achieved by:

- Simple decomposition of ZnO (it is limited to high temperatures ~1400°C)
- Heating Zn powder in the presence of oxygen at 500–700°C (this lowers the process temperatures)
- Heating ZnO powder mixed with graphite at around 1100°C where graphite reduces ZnO to Zn and CO/CO_2 (oxidation occurs later and forms ZnO nanostructures and the presence of graphite lowers the decomposition temperature). The process involves a horizontal tube placed inside a horizontal furnace which contains hot and cold zone at two different ends of the tube. This technique is mainly classified into two types :
  - vapour–solid process (VS) (catalyst free) and
  - vapour–liquid–solid process (VLS) (assisted with catalyst)

However, vapour–solid usually gives a rich variety of nanostructures. Kong et al deposited nanostructures on alumina substrates placed in the cold zone at ~500°C by simple decomposition of ZnO powder in the hot zone at 1350°C. Ren et al grew hierarchical ZnO nanostructures by heating the powder mixture (contains Zno, InO_3 and graphite) at 820 ~870°C. Yao and

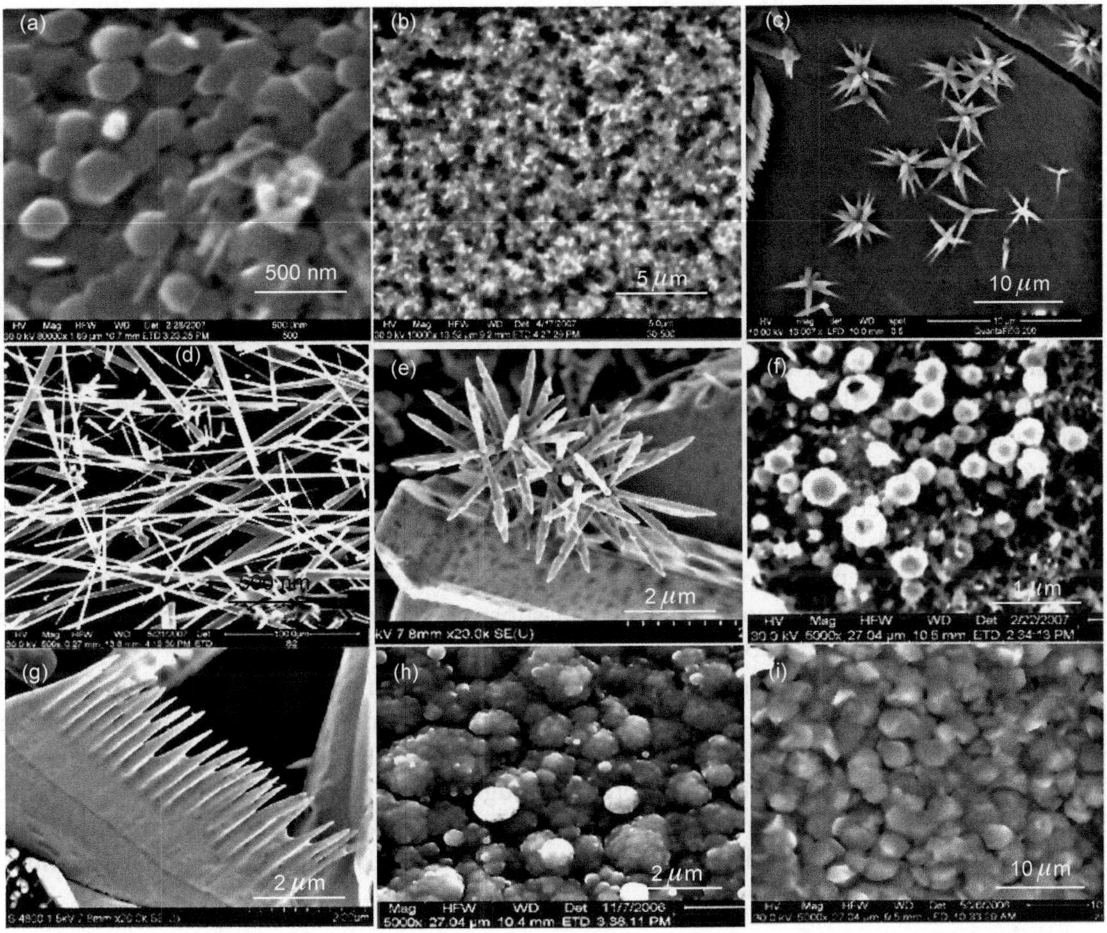

Fig. 6.16 SEM images of ZnO nanostructures synthesized using different techniques. ZnO (a) hexagons, (b) wires and (c) tetrapods by chemical synthesis route (d) ZnO wires, (e) tetrapods, (f) spheres, (g) brush grown by thermal evaporation technique and (h) ZnO spheres and (i) hexagons grown by pulsed laser deposition technique. (*Source:* MS Ramachandra Rao, IIT Madras).

co-workers reported the simplest process which involves heating ZnO and graphite powder mixture at 1100°C; while cooling, they observed the growth of nano-ribbons and nanorods on the furnace walls.

After the invention of carbon nanotubes, ZnO nanotubes and nano-tetrapods have attained much interest. In literature, we have reported the synthesis of ZnO nanotubes by wet oxidation methods which involves the heating of Zn and ZnO powders at ~1300°C in the presence of water. The wall thickness of the tube was seen to be 4–10 nm. ZnO nano-tetrapods were deposited by Wan et al in the VS process with rapid heating of ZnO pellets at 900°C. Even though the

VS process gives a wide range of nanostructures, its lack of control on the alignment, precise location and geometry makes it inferior to the catalyst-assisted VLS process. The VLS process usually uses metal nanoparticles and nanoclusters (e.g. Au, Pt, Cu, Co and Sn) as catalysts to synthesize 1D nanostructures. The molten eutectic alloy of the catalyst with the source material acts as a nucleation site or nucleation seed for nanostructures. The Zn deposited on catalyst particles will react with oxygen vapour and form ZnO which further acts as a nucleation site for nanostructural growth. Dimensions of nanostructures can be varied by changing the size of the catalyst nanoparticles or nanoclusters.

The control over the growth and alignment of ZnO nanorods and nanowires has been achieved by pattern growth. The patterns of different geometries can be prepared by lithography by using a shadow mask during catalyst deposition, or by using TEM grids to get selective deposition with the desired geometry of catalyst. Vertically aligned ZnO rods and wires are finding application as electron field emitters, vertical transistors and UV lasers. Another unique method to grow vertically aligned nanowire arrays involves monolayer self-assembly and VLS nanostructure growth. A monolayer deposit of submicron polystyrene spheres on the substrate which functions as the shadow mask for evaporation of catalyst metal yields a highly ordered array of catalyst spots with desired geometry. Polystyrene spheres are etched out using toulene thereby exposing the metallic substrate underneath. ZnO nanostructures are then grown by the VLS method.

The lattice mismatch with substrate determines the quality of vertically aligned ZnO nanowires. Though sapphire is a widely used epitaxy substrate the ZnO nanowires grown on GaN epilayer give better vertical alignment. Because of its better electrical properties, GaN is preferred in comparison to sapphire. Though vapour transport process has been the dominant deposition technique for nanostructures of ZnO, there are other methods like electrodeposition, sol–gel and polymer-assisted growth which yield nanostructures at much lower temperatures.

6.4.3 Applications of ZnO nanostructures

ZnO nanostructures are the building blocks of new generation electronics and are potential candidates for semiconductors, piezoelectric material, nanosensors, nano-cantilevers, short wavelength light-emitters, field emitters, luminescence devices, gas sensors, UV lasers, solar cells and bio-sensors.

Piezoelectric sensors

The control over flow of current in 1D ZnO nanostructures can be achieved by change in electrical charge distribution during bending. Nanotubes and nanowires developed by piezoelectric materials have a property of charge separation (positive on one side and negative on other) during bending. This fundamental property of nanowires and nano-belts can be used in developing piezotronic sensors to detect force in nano- and pico-Newton range by just measuring the change in current flow through the array of nanostructures; these sensors can be used for measuring the blood pressure in humans. ZnO-based nano-generators are generating significant amount of charge by bending and releasing the array of ZnO nanowires. ZnO nanowires and nanotubes have special advantages which make them potential candidates for an integrated nanosystem. The advantages include:

- Large amounts of deformation without damage. This enables their use in flexible electronics like folding power sources which permit large-volume density of power output.
- Bio-compatibility. This allows usage in the body without any toxic effects.

ZnO nanostructures are also finding application as field effect transistors; the change in conductance occurs by bending the nanostructure between the source and drain which controls the flow of current. Bending produces a gate potential with change in conductance which is directly related to the degree of bending. Current production by bending and releasing of ZnO nanowires could eliminate batteries to supply power to nanoscale systems.

Biosensors

ZnO nanostructures are used as signal enhancing platforms for rapid, multiplexed, high-throughput, high-sensitive DNA sensor arrays. ZnO nanostructures can be effectively used for identification of bio-threat agents by successful discrimination of its DNA sequence from other genetically related species. ZnO nanomaterials are critical for the detection of hybridized DNA, as they achieve increased fluorescence.

Solar cells

Solar cells based on ZnO nano-sheets exhibited better performance than nanoparticle-based solar cells or well-oriented nanowire-based solar cells. ZnO nanosheets have certain advantages including the depression of loss during photoelectron transport, increased dye compound adsorption, and enhanced incident light capture. ZnO nanocrystalline sheets self-assembled on a flexible polymer substrate act as electrodes for dye-sensitized solar cells. This provides an opportunity to integrate solar cells with nano-devices and microcircuits which make it work with self-supplied solar energy. ZnO films with nano-sheet microstructure appear as a matrix of randomly oriented nano-sheets; it combines with a flexible substrate more stably and handles deformation well with increased photon capture.

Gas sensor

ZnO-based structures have attracted great interest as gas sensors because of their high chemical sensitivity to different absorbed gases, amenability towards doping, non-toxic property and low cost. The sensitivity of ZnO nano wires to detect ppm levels of acetone and ethanol has been reported.

6.5 NANOCRYSTALLINE TITANIUM OXIDE

Titania has three allotropic forms—brookite (orthorhombic), anatase (tetragonal) and rutile (tetragonal). Rutile (TiO_2) is the high temperature phase. Anatase and brookite convert to rutile on heating. Nano-titania has found wide application as pigments, absorbents, catalytic supports, gas and humidity sensors and coatings for self-cleaning surfaces. In general, anatase is considered to have higher photocatalytic activity than rutile, while rutile is more suited for white pigments and coatings because of its high refractive index, weatherability (oxidation and wear resistance), and high thermal and chemical stability. It is used in products for its sterilizing, deodorizing and anti-fouling properties. It is also used in the Grätzel cell, a type of chemical solar cell. Nano-titania is used in pigments due to its high refractive index and bright white colour. It is used as an opacifier in glass, cosmetics, sunscreens, paper, paints, etc. It has good resistance to discolouration under UV light. The anatase form of titania is a good photocatalyst under UV light. TiO_2 also shows photocatalytic activity under visible light, provided it is doped with appropriate impurities. Titanium dioxide has the potential to carry out hydrolysis—break water into hydrogen and oxygen—for use in energy production. The efficiency of titanium dioxide as a photocatalyst for this process can be enhanced by doping it with carbon. Nano-titania is used as self-cleaning coatings for disinfecting surfaces under UV radiation. These properties make the material a candidate for application in medical devices, food preparation surfaces, air conditioning filters, and sanitaryware surfaces. Titania is also used as oxygen sensors.

6.5.1 Titania nanopowders

Nanostructured titania in both rutile and anatase forms have been synthesised in a variety of geometrical configurations, like spherical powders, thin films, nanorods, nanowires, etc. The most common methods for synthesis of nano-titania include chemical decomposition methods, sol–gel, chemical vapour deposition, microemulsion and hydrolysis. Of these, the sol–gel route has proven to be a viable technique involving hydrolysis and polycondensation of molecular precursors, such as metal alkoxides, to yield hydroxides or oxides under different conditions.

Nanopowder of anatase titania has been prepared by the sol–gel route using tetrabutyl titanate as precursor. Further calcinations at temperatures of about 800°C have led to transformation of anatase to nano-rutile powder. By using different combinations of precursors, it is possible to directly synthesise rutile nanopowders by sol–gel technique at much lower temperatures. For example, the sol–gel technique was employed to synthesize titania nanopowder, starting with titanium *n*-tetrabutoxide as the precursor. A mixture of catalyst (hydrochloric acid, HCl), water and 50% of the solvent (ethyl alcohol) was added dropwise to a constantly stirred mixture of the precursor and the remaining solvent. Stirring leads to a homogeneous sol, which on drying in air gives a viscous wet gel due to hydrolysis and polycondensation. This gel was oven-dried at 80°C for 10 h to remove most of the water and organic content—categorised as the 'as-prepared' sample. The ratio of precursor:solvent:catalyst:water was taken as 1:15:0.3:x, with $x = 1$–4 and the catalyst being HCl. With the water content at $x = 4$, the influence of changing the catalyst to acetyl acetone was also studied. The as-prepared powders produced were calcined at various temperatures (200–600°C) for 2 h. Figure 6.17 shows a typical SEM micrograph of nanocrystalline titania.

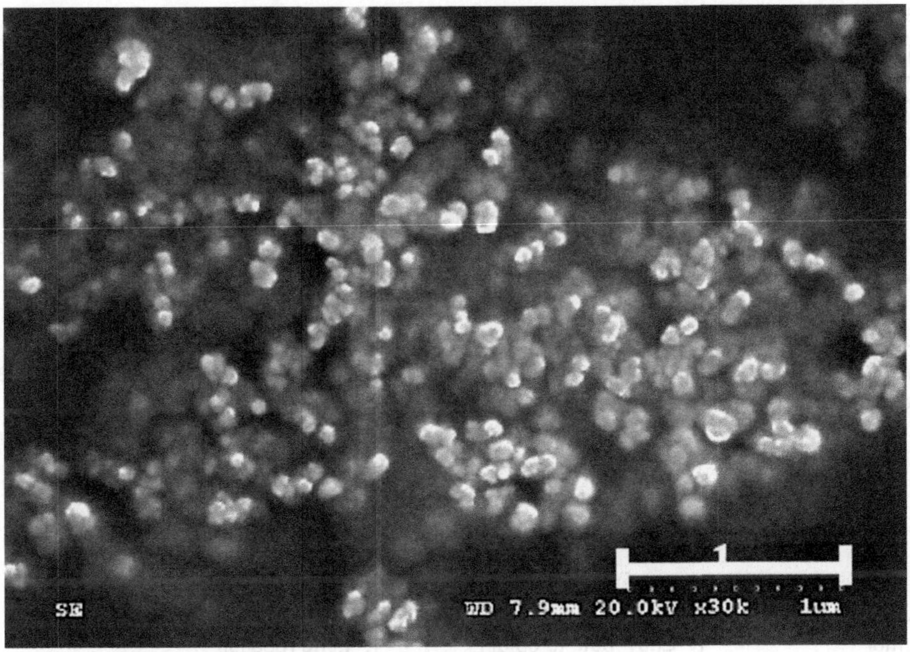

Fig. 6.17 Nanocrystalline titania. (*Source:* BS Murty, IIT Madras)

6.5.2 Titania nanotubes (TNTs)

The increasing prominence of TNT is due to their large specific surface area, photocatalytic potential, and ion-exchangeable ability. Currently, the most common methods of synthesising TNTs include template-based method, hydrothermal treatment and electrochemical anodic oxidation. Electrochemical anodic oxidation involves controlled anodisation of Ti foil to obtain nanoporous titania.

Template-assisted growth of titanium nanotubes is yet another emerging method. Here, nanoporous membranes made of anodic aluminum oxide (AAO) are used as the template. AAO consists of an array of parallel straight nanopores with uniform diameter and length and thus serves as a template for growth of nanotubes. However, the template-assisted method often encounters difficulties in prefabrication and post-removal of the templates and usually results in impurities.

However, for large-scale synthesis of titania for use in engineering applications, the hydrothermal method is the most suitable. The morphology and properties of the TNTs synthesized by hydrothermal technique depend on several experimental parameters like applied temperature, treatment time and type of precursors used. The length of TNTs relates directly to applied temperature (usually a few hundred degrees Celsius).

Comparison of current methods in TNT fabrication

Fabrication method	Advantages	Disadvantages	TNT features
Template-assisted method	1. The scale of nanotube can be moderately controlled by applied template	1. Complicated fabrication process 2. Tube morphology may be destroyed during fabrication process	Ordered arrays (powder form)
Electrochemical anodic oxidation method	1. More desirable for practical applications 2. Ordered alignment with high aspect ratio 3. Feasible for extensive applications	1. Mass production is limited 2. Rapid formation kinetics is subjected to the utilisation of HF 3. High expense of fabrication apparatus	Oriented arrays (thin film)
Hydrothermal treatment	1. Easy route to obtain nanotube morphology 2. A number of modifications can be used to enhance the altributes of titanium nanotubes 3. Feasible for extensive applications	1. Long reaction duration is needed 2. Highly concentrated NaOH must be added 3. Difficulty in achieving uniform size	Random alignment (powder form)

6.6 MULTILAYERED FILMS

Multilayered materials with individual layers that are a micron or less in thickness, sometimes referred to as super-lattices, can exhibit substantial enhancement in hardness or strength. The enhancement in hardness can be as much as 100% when compared to the value expected from the rule of mixtures, which is essentially a weighted average of the hardness of the constituents of the two layers. Table 6.1 shows how the properties of isostructural multilayers can show a substantial increase in hardness over that for fully interdiffused layers. The table also shows how there can be a substantial enhancement in hardness for non-isostructural multilayers compared to the values for the same materials when they are homogeneous.

There are many factors that contribute to enhanced hardness in multilayers. These can be summarised as:

1. Hall–Petch behaviour
2. Orowan strengthening

Table 6.1 Results for some experimental studies on multilayer hardness

Study	Multilayer	Maximum hardness and multilayer repeat length	Reference hardness value	Range of hardness value for multilayers
Isostructural Knoop hardness	Cu/Ni	524 at 11.6 nm	284 (interdiffused)	295–524
Non-isostructural nanoindentation	Mo/NbN	33GPa at 2 nm	NbN– 7GPa Mo–2.7GPa	12–33GPa
—	W/NbN	29GPa at 3 nm	W–7Gpa	23–29GPa

3. Image effects
4. Coherency and thermal stresses
5. Composition modulation

Hall–Petch behaviour

It is related to dislocations *piling up* at grain boundaries. (Pile-up is used to describe two distinct effects: one is material building up at the side of an indentation, the other is an accumulation of dislocations on a slip-plane.) The dislocation pile-up at grain boundaries impedes the motion of dislocations. For materials with a fine grain structure, there are many grain boundaries, and, hence, dislocations find it hard to move. In polycrystalline multilayers, it is often the case that the size of the grains within a layer scales with the layer thickness so that reducing the layer thickness reduces the grain size. Thus, the Hall–Petch relationship (below) should be applicable to polycrystalline multilayer films with the grain size, d_g, replaced by the layer thickness.

$$Y = Y_o + k_{HP} d_g^{-0.5}$$

where Y is the enhanced yield stress, Y_o is the yield stress for a single crystal, and k_{HP} is a constant.

There is an ongoing argument about whether Hall–Petch behaviour really takes place in nanostructured multilayers. The basic model assumes many dislocations are present in the pile-up, but such large dislocation pile-ups are not seen in small grains and are unlikely to be present in multilayers. As a direct consequence, studies have found a range of values, between 0 and −1, for the exponent in equation, rather than the – 0.5 predicted for Hall–Petch behaviour.

Orowan strengthening

This is due to dislocations in layered materials being effectively pinned at the interfaces. As a result, the dislocations are forced to *bow out* along the layers. In narrow films, dislocations are pinned at both the top and bottom interfaces of a layer and bow out parallel to the plane of the interface. Forcing a dislocation to bow out in a layered material requires an increase in the applied shear stress beyond that required to bow out a dislocation in a homogeneous sample. This additional shear stress would be expected to increase as the film thickness is reduced.

Image effects

These were suggested by Koehler as a possible source of enhanced yield stress in multilayered materials. If two metals, A and B, are used to make a laminate and one of them, A, has high dislocation line energy, but the other, B, has low dislocation line energy, then there will be an increased resistance to dislocation motion due to image forces. However, if the individual layers are thick enough that there may be a dislocation source present within the layer, then dislocations could pile-up at the interface. This will create a local stress concentration point and the enhancement to strength will be very limited. If the layers are thin enough to have no dislocation source present, the enhanced mechanical strength may be substantial. In Koehler's model only nearest neighbour layers were taken to contribute to the image forces. However, this was extended to include more layers without substantial changes in the results. The consequence on image effects of reducing the thickness of the individual layers in a multilayer is that it prevents dislocation sources from being active within the layer.

Coherency and thermal stresses

For many multilayer systems, there is an increase in strength as the bilayer repeat length is reduced, but there is often a *critical repeat length* (e.g., 3 nm for the W/NbN multilayer of Table 6.1) below which the strength falls. One explanation for the fall in strength involves the effects of coherency and thermal stress on dislocation energy. Unlike image effects where the energy of dislocations are maximum or minimum in the centre of layers, the energy maxima and minima are at the interfaces for coherency stresses. Combining the effects of varying moduli coherency stresses shows that the dependence of strength on layer thickness has a peak near the repeat period where coherency strains begin decreasing.

Composition modulation

Another source of deviation in behaviour at small repeat periods is the imperfect nature of interfaces. With the exception of atomically perfect epitaxial films, interfaces are generally not atomically flat and there is some interdiffusion. For the Cu/Ni film of Table 6.1, the effects of interdiffusion on hardness were examined by annealing the multilayers. The results were in agreement with a model by Krzanowski that predicted the variations in hardness would be proportional to the amplitude of the composition modulation.

It is interesting to note that the explanations for enhanced mechanical properties in multilayered materials are all based on dislocation mechanisms. So it would seem natural to assume that multilayered materials that do not contain dislocations will show no enhanced hardness over their *rule of mixtures values*. This has been verified by studies on amorphous metal multilayers, which shows that the hardness of the multilayers, firstly, lies between that of the two individual materials and, secondly, has almost no variation with repeat period.

SUMMARY

- Advance in the fabrication technologies of nano electronic structures has led to the realization of devices like single electron transistor, field effect quantum dot etc., for a wide range of end applications.
- The properties of CNT depend to a large extent on its composition, chirality and boundary structure, which can in turn be controlled by adjusting the growth parameters.
- Wide band gap nanostructured semiconductors based on GaN and ZnO have been extensively employed as sensors, microelectronic devices and for other functional applications.
- Mechanisms of high hardness of multilayered nanostructured thin films have been discussed.

EXERCISES

1. What is the basis of quantum confinement in quantum dots?
2. What controls the properties of quantum dots?
3. What are the most popular applications of quantum dots?
4. What do you understand by chirality of carbon nanotubes? How can it be controlled and how does it affect the properties of nanotubes?
5. What are the popular techniques used for the synthesis of carbon nanotubes?
6. What are the purification methods for carbon nanotubes?
7. What makes carbon nanotubes suitable for sensor applications?
8. What makes nanowires interesting? What are their applications?
9. What are the most important applications of nanocrystyalline ZnO? Explain the principle adopted in these applications.
10. What is the principle behind the self-cleaning behaviour of nanocrystalline titania?

Chapter 7

Concerns and Challenges of Nanotechnology

Learning objectives

- Environmental, ecological and health hazards of nanoparticles
- Nanotoxicology and its effect

One of the most oft quoted but extremely important sayings can be traced to the late physicist Richard A Feynmann. The expression "*There is plenty of room at the bottom*", captured the minds of generations of scientists and triggered a whole new science and revolutionary technology. Nearly five decades after Feynman's lecture, nanotechnology enhanced products are increasingly used in routine as well as high-end cutting-edge technology applications. More exciting possibilities exist in biomedical, energy and environmental related applications. Nano-engineered materials have witnessed extensive application in pollution control, purification and desalination of water and in effective waste management of hazardous by-products. It is a popular belief that the nano-revolution is set to have a far larger global econo-techno-political impact than the industrial revolution of the nineteenth century or the information technology revolution of the twentieth century.

There are indications that nanotechnology has the power to repair the brain. Fundamental properties of carbon nanotubes, such as higher thermal conductivity, are being used for making faster electrical signal conductors and to form intimate mechanical contacts with cellular membranes, thereby establishing a functional link to neuronal structures, which in turn can dramatically increase the speed of the brain.

Nanomaterials characterized by widely different defect dynamics have unique structural and functional properties in comparison to bulk materials. The high surface area-to-volume ratio of nanomaterials results in their higher chemical and biological activity as the surface atoms are unsaturated in their chemical bonds. With nearly a quarter or more atoms residing at the surface, nanomaterials can be exceptionally reactive. For example, using nanoscale silver fibres for water-purification filters has greatly enhanced the effectiveness of such filters.

Tiny science, huge concern

Nanotechnology can be characterized as passive or active. Passive applications are those in which the nanomaterial or its structure does not change form or function. For example, there are numerous cosmetics and silver-based anti-microbial products (including food containers and no-smell socks) and other products, ranging from tennis racquets to teas, that use nanomaterials.

Nanostructures or nanomaterials are said to be active when they are able to change their form or function. A simple example is an anti-cancer drug in which a dendrimer is designed to find cancer cells and then release a chemical that kills them. The increasing overlap of biotechnology and nanotechnology will eventually lead to nanosystems being used as robots.

Does nanotechnology pose health risks?

As science and technology develop and advance, the environment and ecological systems are at great risk, as there is a deviation from natural forces of equilibrium. From the primitive invention of fire by pre-historic humans to the development of advanced air transport systems, it has been seen that these technologies can also be employed for societal disturbances in the hands/minds of unethical groups of people. There are similar problems associated with nanotechnology.

Some nano-fabrication methods use toxic raw materials or produce toxic by-products (for example, some carbon nanotube synthesis routes). This needs to be fully investigated to understand the extent of the harmful effects of engineered nanoparticles, as the degree of influence is species specific and often depends on the size and geometry of the particle. Extensive research is underway to characterize the nanotoxicological effect of different nanoparticles on aquatic and animal species.

As discussed in Chapter 1, nanoparticles have a unique set of properties (physical, chemical and mechanical), widely different from their own bulk counterparts. A good review on the adverse effects of nanomaterials has been provided by Dr Fadeel of Karolinska Institute's Division of Biochemical Toxicology, who has summarized the proceedings at the Stockholm Symposium on Nanotoxicology (There's plenty of room at the forum: Potential risks and safety assessment of engineered nanomaterials, *NanoToxicology*, 2007). Extensive cross-disciplinary collaborations are therefore required to quantify and evaluate the risk assessment in detail. This will finally lead us to the goal of safe handling of nanotechnologies.

Once the safe limits of different nanomaterials are evaluated and tabulated, measures to mitigate the risks involved in undesirable exposure during manufacturing or service exploitation of nanomaterials can be achieved. There are also concerns on waste management of nano-enhanced products and their contribution to environmental pollution. There is a need for quantitative data followed by intensive scientific insight into the risks to human beings and the environment due to genetically modified organisms (GMOs) in agriculture and service applications.

As a result of these concerns, a new field of research termed as nano (eco-)toxicology has emerged in the last decade. This field studies the effect of engineered nanoparticles on living organisms. In the following sections, we will highlight the health and environmental issues related to the use of engineered nanomaterials and try to draw a clear picture of happenings to mitigate such risks.

What makes it dangerous?

The adverse effects of certain chemical substances, even in the bulk form, have been well studied for several centuries. Paracelsus (1493–1541), widely regarded as the father of toxicology, expressed the view that most materials are toxic when their quantity in the human system exceeds a *critical acceptable limit*. The maximum dose or acceptable concentration of various materials without toxic effects is therefore of paramount importance, particularly in the field of nanomaterials. In nanotoxicology, the critical question is which particle characteristics are crucial in initiating and causing adverse effects? It is known that the higher surface area-to-volume ratio of nanoparticles enhances their chemical and bio-reactivity, as demonstrated by pioneering animal studies of particle deposition and retention in the lung. It is this combined chemical and bio-reactivity that makes the risk factor of selective nanoparticles even higher. Certain other foreign bodies and toxic materials can adhere to chemically active nanomaterial sites and further enhance the ill effects as the nanomaterials react with biological systems. The risk involved is that nanomaterials can not only penetrate cell membranes allowing entry of otherwise restricted foreign bodies, but they can also translocate to other tissues and organs. Thus, detailed investigations to understand the fundamental mechanisms are vital to curb the adverse effects of engineered nanoparticles.

Potential impact of nanomaterials on humans and the environment

Due to the increasing use of synthetic nanoparticles, the concentration of such particles in environmental media, i.e., soil, water and air, is expected to increase in the future. Similar to larger airborne particulates, nanoparticles form during technological as well as natural processes, such as volcanic eruptions or forest fires. Naturally produced nanoparticles vary considerably in their shape, composition and size. In contrast, manufactured or 'deliberately' produced nanoparticles are of uniform design to comply with the desired properties. Studies on naturally formed ultrafine particulates or those produced by combustion permit estimates concerning the behaviour and impact of the nanoparticles on the environment. However, these estimates are insufficient to reliably assess the risks carried by manufactured nanoparticles. The broad applicability of nanotechnology and the great differences existing between the various nanomaterials requires a differentiated evaluation of their potential risks.

As nanotechnology is still at its infancy, most theories on nano–bio interactions have been speculative and need to be validated by controlled experiments. Detailed investigation on the effect of nanoparticles on aquatic, plant and animal species are being undertaken, as seen from a large number of publications. However, significant efforts are needed in this direction before regulatory measures can be introduced. Figure 7.1 shows the nano–bio interactions.

Cytotoxicity is the quality of being toxic to cells. Treating cells with a cytotoxic compound can result in a variety of cell fates. The cells may undergo necrosis, in which case they lose membrane integrity and die rapidly as a result of *cell lysis*. The cells can stop actively growing

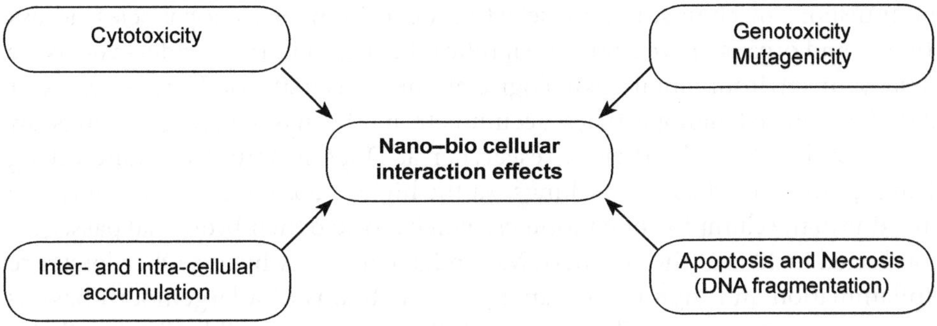

Fig. 7.1 Nano–bio interactions.

and dividing (a decrease in cell viability), or the cells can activate a genetic programme of controlled cell death (*apoptosis*). *Genotoxicity* describes a deleterious action on a cell's genetic material affecting its integrity.

The ill effects of carbon nanotubes on the lungs of animals have been reported by several investigators. There are suggestions that some nanotubes can be as harmful as asbestos if inhaled. Studies in rats have shown that when inhaled, some nanomaterials can be transported from the respiratory tract and deposited in brain cells, affecting neural junctions. There is also emerging evidence that certain nanomaterials can interfere with proteins and DNA in the human body, disrupting the work of the building blocks of life. Because of their large surface-to-mass ratio, nanoparticles are more reactive than ordinary materials.

They can also be far more explosive than ordinary-sized materials—a property that the military is actively exploiting. In 2007, the Russians appear to have exploded the first bomb based on nanotechnology. It was reportedly the largest non-nuclear bomb ever tested.

Nanomaterials have been used experimentally to improve the environment by cleaning up waste sites, purifying water and filtering air. Many applications of nanomaterials replace toxic substances or substitute more energy-intensive processes. Both the good and deleterious environmental effects of nanomaterials are largely unknown. On the negative side, nanomaterials will be extremely difficult or impossible to remove from the environment. Nano-silver is an example of the Jekyll-and-Hyde character of nanomaterials. The anti-microbial properties of nano-silver are being used in a variety of products, but the same properties may pose a threat to the environment as well as to waste-treatment plants that clean sewage through bacterial action.

Health issues with engineered nanomaterials

There has been little research in the potential health risks of nanotechnology-based products that are already available in the market, for example, cosmetics. Although the risks of exposure to nanomaterials are just being reported, an article in *Science* (2006) by Nie and his group from the University of California, Los Angeles, highlights the possible damage that nanomaterials can inflict on living organisms by means of cellular and subcellular interactions. The vastly different transport mechanisms and selective deposition of nanoparticles across various zones, tissues and organs are reasons for significant concern in their wide-scale usage. Gunter Oberdorster, an environmental toxicologist at the University of Rochester, has indicated (in 2004) that inhaled nanoparticles accumulate in the nasal passages, lungs and brains of rats. A study by David Warheit, a researcher at Dupont, found immune cells gathering around clumps of nanotubes in rats' lungs. At the highest dose used in the study, the animals suffocated due to the clumping of nanotubes, which blocked their bronchial passages. Another study conducted in 2004 at the Southern Methodist University by Eva Oberdorster reportedly found inflammation and 'significant damage' in the brains of a largemouth bass as a result of exposure to nanomaterials called aqueous fullerenes. The possible diseases that can affect different parts of the body owing to nanoparticles are listed in Table 7.1.

Table 7.1 Possible diseases that can affect different parts of the body owing to nanoparticles

Brain	Alzheimers & Parkinson's disease
Lungs	Cancer, asthma, bronchitis
Gastrointestinal system	Colon cancer
Circulation	Blood pressure, thrombus, vasoconstriction
Skin	Dermatitis, auto-immune disease
Lymphatic system	Sarcoma

There is something fishy about these nanotubes

Five different cases studying the inhalation and dermatological risks associated with carbon nanotubes and ultrafine particles are given below. Three of these studies were presented at a recent conference of the American Chemical Society (ACS). The first series of studies is concerned with the toxicity of ultrafine particles.

Dr Gunter Oberdorster at the University of Rochester has a great deal of experience in the field of inhalation and dermatological risks posed by the ultrafine particles (UFP) family (<0.1 μm). Nanoparticles fall under this realm. In order to answer the difficult question, whether nanoparticles pose potential toxicity, Prof Gunter used nanocarbon particles for his various studies. Interestingly, the nanocarbon particles penetrated deeper into the lungs as the time of exposure increased. His research articles on this subject show that there is a greater possibility for the blood–brain barrier to be breached, causing a huge impact on the central nervous system.

Dr CW Lam at Wyle Labs, NASA Johnson Space Center, and Dr Robert Hunter, University of Texas (Houston), studied the impact of carbon nanotubes on the lungs. The nanocarbon particles were suspended in a fluid which was in direct contact with the lungs of mice. Hunter found that the nanotubes clumped together into bundles and stimulated an immune response, which left scar tissue on the lungs. In the conclusion, Hunter urges: "People should really take precautions. Nanotubes can be highly toxic."

Dr David Warheit at DuPont's Haskell Labs instilled a single-walled carbon nanotube soot mixture into the trachea of rats. For comparative purposes, he also instilled a group with carbonyl iron and quartz, respectively. Fifteen per cent of the rats treated with carbon nanotubes suffocated to death within twenty-four hours due to clumping of the nanotubes that obstructed the bronchial passageways. The quartz-instilled and carbonyl iron-instilled rats had some toxicity and no toxicity, respectively. The main conclusion of this particular study was that carbon nanotubes might be irrespirable.

A few years ago, at the University of Warsaw, two studies released at the same time studied the dermatological and inhalation effects of carbon nanotubes. In the study on dermatological effects, which used rabbits, researchers were surprised to not find any signs of health hazards related to skin irritation and allergic risks, in contrast to the general assumption that nanotubes cause health hazards. The study recommended no special precautions with respect to carbon nanotubes in the working environment; in fact, the article was titled 'Carbon Nanotubes: Experimental Evidence for a Null Risk of Skin Irritation and Allergy'. In the next study, 'Physiological Testing of Carbon Nanotubes: Are They Asbestos-like?' the researchers found that carbon nanotubes do not exhibit effects similar to asbestos. In this study, a carbon nanotube soot mixture was instilled into the tracheas of guinea pigs and it was found that this mixture is not associated with any health risks.

Unfortunately, these studies show contradicting results about the toxicity of carbon nanotubes. They do not deal with the associated issue of exposure. A recent collaborative study by the National Institute for Occupational Safety and Health (NIOSH), NASA, Rice University, and Carbon Nanotechnologies Inc., emphasized that worker exposure to carbon nanotubes would be low at low agitation levels, but urged caution until more is known about toxicity. This obviously points to the important relationship between toxicity and exposure in determining the overall risk to humans and the environment. The Office of Research and Development at the EPA has requested nanotoxicologists to conduct detailed studies on the environmental effects of nanotechnology, and suggested that this time the studies should be independent, inhalation studies. An apt place to start may be with workers at factories that produce significant amounts of carbon nanotubes or other nanoscale substances for either research or commercial purposes.

Adverse effects of quantum dots

Among the large array of nanomaterials, quantum dots (QD) are of particular importance for their possible therapeutic and diagnostic medical applications. A discussion on the toxicity of QDs can be confusing, primarily because of their diverse synthesizing technologies—materials synthesized by one technology will be highly toxic, while those synthesized by another may not be that toxic.

A summary of the studies on quantum dots is given below:

- The toxic effect of quantum dots depends on multiple factors, including their physicochemical properties and environmental conditions.
- Exposure of human beings to QDs might be from the environment or during therapeutic use.

Since QDs can accumulate within organs and tissues (e.g. in lungs), there is a need for extensive studies to explore their health risks.

- Introduction of QD into the environment may occur via waste streams from industries, research and clinical settings. The persistence of these materials in the environment may be of long duration and the environmental exposure will depend on the partition of QDs between water, air and various soil types.
- The main concern of QDs is their content in metals, such as cadmium or selenium, that are known to cause adverse effects on the environment and in vertebrates, including humans.

Consequently, it is the stability of the metalloid core-coating complexes that may render QD potentially harmful. In particular, the oxidative, photolytic and mechanical stability of the QD coating and metalloid core are the most important parameters governing the bio-reactivity and risk concerns of QD.

As has been emphasized, there are several parameters that decide the risk quotient of QD exposure. For example, some QDs have been found to be cytotoxic only after oxidative and photolytic degradation of their core coatings. Lastly, it is difficult to make statistical conclusions from present literature resources, largely due to the varying units of measurement used and insufficient data/characterization of all other parameters that could have influenced the bio-activity of the species. Hence, the need for wide-scale cross-disciplinary and cross-national networking is of paramount importance in understanding nanotoxicological effects.

Nanoparticles in living systems

All nanoparticles, on exposure to tissues and fluids of the body, have a tendency to be adsorbed to macromolecules at the site of entry. The extent of adsorption and further mechanisms of cellular interaction will depend on several aspects, including surface chemistry and geometry. The specific features of this adsorption process will depend on the surface characteristics and geometrical features of the particles. By anchoring specific bio-molecular linkers to nanomaterials, their site-specific reactivity can be characterized. Such studies can be employed for optical labelling of biomolecules and for targetted delivery of drugs. There can be a number of areas of possible human exposure during the production of nanoparticles.

Respiratory tract Classic routes of exposure of the body are through the respiratory tract by inhalation, through the skin by absorption and through the mouth by ingestion, or a combination of these routes. The most important route of exposure is probably that of inhalation through the respiratory tract. The number of research activities referring to the impact of nanoparticles has been low so far. However, most of the available scientific studies refer to inhalation by the respiratory route. In contrast to larger particles, nanoparticles can reach the alveolar region of the lungs. Because of their small size, they are removed insufficiently by the alveolar macrophages. This may cause inflammation in the lungs.

In a review article for ECTOC that appeared in *Particle and Fibre Toxicology,* the effects of nanoparticles inside the lungs was examined on the basis of the dissolution behaviour of nanoparticles in mucus and also based on the interaction of nanoparticles with epithelium cells in the respiratory tract. It was reported that nanoparticles that are either insoluble or dissolve slowly in the mucus are eventually moved, by the action of ciliated cells, to the throat, where they are swallowed. The smaller nanoparticles may be retained within the tissues and organs in the respiratory tract.

The few nanoparticles that were deposited in the alveolar region were digested by alveolar macrophages and men eventually passed out of the body through the gastrointestinal tract. Therefore, the extent of interaction and side effects of such particles depend more on interaction with alveolar macrophages. The remaining nanoparticles may be translocated to other body cells by intracellular and transcellular migration. Hence, insoluble nanoparticles may linger for months and years in the lungs and other body tissues and organs. It has also been demonstrated that some of the nanoparticles can enter the bloodstream and the brain tissues after being inhaled through the respiratory tract. Studies in rats have demonstrated that direct absorption of particles through the nose and into the brain is possible. However, possible adverse effects have not yet been sufficiently examined.

Dermal absorption With regard to dermal absorption, two pathways are conceivable: particles can enter the skin through the upper layer of the skin or through the hair roots. However, there is insufficient knowledge regarding intact and pre-damaged skin. It also remains unclear whether toxic substances adhering to the particles may enter the body by the dermal route of exposure. Investigations into the importance of this route of exposure should be given priority since nanoparticles are already present in a great number of skincare products. In addition to the deliberate uptake of nanoparticles through the mouth (such as ingredients of pharmaceutical products), unintentional oral intake (such as those contained in certain food items) should also be taken care of.

Intestines In principle, insoluble particles can be absorbed through the intestine and thus enter the lymphatic system. From there, such particles may enter blood circulation and spread throughout the body. However, no studies are available at present to provide sufficient information about the assessment of possible risks involved in the ingestion of nanoparticles.

Bloodstream The relative importance of the individual routes of exposure cannot be described in quantitative terms at present. It is, however, understood that particles entering the body via the bloodstream may be transported to a number of organs (heart, liver, spleen, kidneys and bone marrow). Studies have indicated that nanoparticles are capable of crossing biological barriers such as the blood–brain barrier. Also, it has to be assumed that a transfer of nanoparticles through the placenta into the foetus is possible. Irrespective of the chances of a specific utilization of these mechanisms for therapeutic purposes, this may also involve risks.

Cells At the cellular level, barriers such as cell membranes do not constitute obstacles for nanoparticles. In nerve cells, particles were observed to move along the axis cylinders. A great number of interactions with cell components is conceivable for particles penetrating into a cell. However, the health implications of such possible interactions are still unknown. The distribution of nanoparticles in the body seems to depend on their size, form and substance properties. Biodegradable nanoparticles are not a problem as they can be metabolized and excreted. However, non-bio degradable nanoparticles are a major concern. It has to be assumed that accumulation will take place predominantly in the organs of detoxification. Whether or not a risk may arise from such accumulation has not yet been sufficiently examined.

Ecological aspects

The environmental issues of nanotechnology are currently an increasingly important area of research. There are still only few scientific studies available on the impact of nanoparticles on the environment. However, it has to be assumed that due to their special properties, nanoparticles will definitely pose a risk to the environment.

Water fleas (*Daphnia*) may die at relatively low concentrations of C_{60} molecules (buckminster fullerenes) and nanoscale titanium dioxide in the water, depending on the type of administration. Experiments performed in young largemouth bass (*Micropterus salmoides*) have revealed that C_{60} nanoparticles are absorbed through their gills. They overcome the blood–brain barrier and cause damage to the brain, even at low concentrations. In zebrafish (*Danio rerio*) embryos, carbon nanotubes have been shown to cause a delay in the hatching of offspring. Further, the bactericidal activity of some nanomaterials could produce adverse effects in sewage treatment works and cause a change in the composition of the microbial population in water.

Likewise, studies on the impact of nanoparticles on soil ecosystems are almost completely missing. As far as mammals are concerned, the results of laboratory studies modelling the impact on human health can also be extrapolated to wild animals. So far, no studies that refer to other vertebrates or invertebrates are available. Experiments using aluminium nanoparticles have revealed a reduced growth of the roots of a number of crops. This effect was not seen for larger aluminium particles. Due to their biocidal action, i.e., their ability to kill bacteria, nanoparticles may interfere with the composition of the microbial population in the soil.

SUMMARY

- Nanotechnology is bound to have an impact due to large-scale applications in industry and household products.
- Currently, conclusive information on possible adverse health effects on exposure to different types of nanomaterials is not available.
- There is a need to establish regulatory bodies for establishing guidelines and safe limits for ensuring adoption of nanotechnology without causing environmental damage or ecological imbalance.
- In order to substantiate their positive potential, nanotechnological methods and products have to be evaluated for advantages in comparison to conventional alternatives. For example, the Federal Environment Agency supports inverse nanotechnology for the treatment of sewage and water for human consumption.
- The risks associated with nanotechnology have to be studied. There is an urgent need to identify and assess possible risks for human health and the environment due to nanotechnology.

- Despite increasing numbers of scientific studies, there are considerable gaps as far as information is concerned; hence, there is great need for further research.

EXERCISES

1. What is nanotoxicology?
2. What are the possible adverse effects of nanotubes on heath and the environment?

Glossary of Terms

Aerogel: a porous solid formed from a gel in which the liquid is replaced with a gas with gas entrapment

Atomic force microscopy (AFM) or scanning probe microscopy (SPM): a high-resolution device used to map topography or other functional properties of the surface atoms at atomic resolution capabilities

Atomic manipulation: atom by atom modification of surface structure or chemistry made possible by advanced techniques like atomic force microscope and scanning tunnelling microscope

Band gap: energy gap between the valence band and conduction band in a solid in which all electronic energy states are forbidden

Biocompatibility: capability of a material in contact with a biological system to perform its intended function without causing deleterious changes

Biomimetic: the science of imitating or reverse engineering from natural systems to the study and design of engineered systems using modern technology

Bot: a robot or automated intelligent machine

Bottom-up: a strategy for synthesizing nanomaterials from atomic scale fundamental units where the fundamental units link up to form nanoparticles/nanostructures

Buckminster fullerene: a spherical molecule with the formula C_{60}, named in homage to Richard Buckminster Fuller, due to its resemblance to the geodesic dome designed by him; Buckminster fullerene is the first fullerene molecule to be discovered and is also the most common in terms of natural occurrence, as it can be found in small quantities in soot

Catalyst: a substance that does not chemically take part in the reaction but results in increase in rate of reaction by decreasing the activation energy

Charge-coupled device (CCD): a device that can gather position-sensitive charge information and convert to digital data for manipulation, which is used extensively for digital imaging applications

Cell lysis: breaking down of a cell, often by viral, enzymic or osmotic mechanisms that compromise its integrity

Chirality: a crystallographic arrangement of atoms in a molecule such that there is no internal symmetry plane; it will give a non-superimposable mirror image

Complementary metal–oxide semiconductor (CMOS): an emerging technology for the fabrication of ICs and VLSI, the main advantage being low power consumption and high noise, enabling larger density of devices within unit area

Carbon nanotube (CNT): an allotrope of carbon with cylindrical nanostructure and having high aspect ratios; their unusual electronic and magnetic properties find wide applications

Colloid: a homogenous suspension of a dispersoid in a continuous medium; it may be a solid, liquid or gas

Chemical vapour deposition (CVD): a technique for depositing thin films on a substrate using gaseous reactants

Cytotoxicity: the quality of an external chemical or system being toxic to cells; the cells may undergo necrosis, lose membrane integrity and die rapidly as a result of cell lysis

Depletion zone: a region at the junction of semiconducting materials that is devoid of free charge carriers

Dislocation: a crystallographic line defect involving irregularity in the periodic arrangement of atoms (missing row of atoms in a plane) in a crystal

DNA chip: a sensor based on a semiconductor microchip used to identify mutations or alterations in a gene

Equal channel angular pressing (ECAP): a severe plastic deformation technique for producing ultrafine grain structures, which introduces a large amount of shear strain into the materials without changing its shape or dimensions; equichannel angular extrusion (ECAE) is a similar process involving extrusion

Elastic modulus: a material's tendency to resist elastic deformation; it is the slope of the stress–strain tensile curve of a material in the elastic regime

Electron microscope: a microscope that focusses a collimated accelerated electron beam on the specimen to produce a magnified image at atomic resolution

Electronic nose: a device consisting of an array of chemical sensors to detect odours or flavours

Electronic tongue: a device consisting of an array of chemical sensors to detect and compare tastes

Epitaxy: growth of a secondary phase maintaining a perfect crystallographic registry (coherency) with the underlying substrate

Fab: a microfabrication facility consisting of clean rooms and controlled deposition process for the fabrication of semiconductor devices and ICs

Field effect transistor (FET): a transistor whose conductivity can be controlled by electrical field

Fuel cell: an electrochemical cell capable of producing electrical energy with fuel or reactant being used up from an external source

Genetically modified organism (GMO): an organism whose genetic sequence has been modified using genetic engineering techniques such as recombinant DNA technology

Genotoxicity: quality of potentially mutagenic or carcinogenic chemicals to destroy genetic material in an organism

Giant magnetoresistance (GMR): quantum mechanical effect observed in thin film structures: the electrical resistance decreases significantly when the ferromagnetic layer is exposed to a magnetic field

Global positioning system (GPS): a system that provides geographic location and time anywhere on earth

Grain boundary: a 2D defect, the interface bordering two well-defined crystals

Grain boundary migration: coordinated movement of grain boundaries activated either thermally or by mechanical stress

Hall–Petch relation: the effect describing the inverse effect of grain size on the hardness of a crystalline solid that arises mainly due to grain boundary strengthening

Hot isostatic pressing (HIPing): the process of using high hydrostatic pressure and temperature to compress fine particles into coherent parts

Hydrophilic: a molecule that is attracted to water

Hydrophobic: the physical property of a molecule to repel water

In vitro: an experiment carried out in a test tube or controlled environment using synthetically generated body fluids

In vivo: experimentation using living organisms

Kryder's law: the memory storage capacity of hard drives doubles almost every year

Light-emitting diode (LED): a semiconductor light source working on the principle of electroluminescence, where the wavelength of light emitted depends on the band gap of semiconductors

Liquid crystal (LC): a state of matter with properties between a liquid and solid crystal; it is used extensively in liquid crystal displays

Magic number: a critical number of atoms in a cluster size providing it higher structural and potential stability

Mechanical alloying: a solid state process in which grain refinement occurs by repeated deformation, fracturing and cold welding of powder particles in a high-energy ball mill

Melting point oscillation: the phenomenon of suppression of melting point followed by elevation as the particle size is reduced from bulk to sub-nanometre size

Microelectromechanical systems (MEMS): a microdimensional mechanical system driven by electrical energy; when the dimensions of the mechanical devices approach nanometric range they are termed nanoelectromechanical systems (NEMS)

Mesoporous: porous materials with regularly arranged, uniform mesopores (2–50 nm in diameter); their large surface areas make them useful as adsorbents or catalysts

Microcantilever: a cantilever beam with dimensions in the micrometer scale that is extensively used in the field of MEMS, sensors, resonators, etc.

Microprocessor: a CPU integrated inside an electronic device; it is programmable and processes the input data complimented with the instructions stored in its memory and accordingly gives an output

Molecular electronics: the study and application of molecules for electronic device applications

Moore's law: a long-term trend in computing hardware suggesting that the number of transistors built in a unit area of the device approximately doubles every 18 months

Metal–oxide semiconductor field effect transistor (MOSFET): a transistor in which a voltage on the oxide-insulated gate electrode can induce a conducting channel between the two other contacts called source and drain

Multiply twinned particles (MTP): observed frequently with a pseudo five-fold symmetry in nanocrystalline particles and thin films (deposited on crystalline substrates) of cubic face-centred metals, diamond-type semiconductors (C, Si, Ge) and alloys

Multilayers: thin films of differing chemistry or structure deposited one over the other

Nano: Greek prefix meaning dwarf or something very small; depicts one billionth (10^{-9}) of a unit

Nanobots: a robot (semi- or fully-automated intelligent machine) consisting of components of a few hundred nanometre-dimensions; they are also referred to as nanorobots, nanoids, nanites, nanomachines or nanomites

Nanofibre: fibres with diameter less than 100 nm

Nanofluid: colloidal suspension of nanoparticles of metals, ceramic, carbon nanotubes, etc.

Nanoindentation: an indentation hardness test applied to nanoscale volumes at small loads to obtain the hardness of individual nanoparticles

Nanolithography: a nanofabrication technique for patterning nanoscale features; used extensively in the fabrication of ICs and NEMS

Nanomaterial: class of materials in which at least one of the dimensions is on the nanoscale (<100 nm)

Nanoperm alloys: alloys based on the Fe–Zr–B system; contain larger concentrations of Fe and have higher saturation induction compared to the Finemet alloys

Nanorods: 3D nanostructures with aspect ratio typically in the range of 3–5; all their dimensions are in the range 1–100 nm

Nanoshells: a thin coating over a core object a few tens of nanometres in diameter

Nanotechnology: study of manipulating matter on an atomic and molecular scale; generally deals with structures sized between 1 and 100 nanometres in at least one dimension, and involves developing materials or devices possessing at least one dimension within that size

Nanowires: 1D nanostructures with width of nanometric dimensions and exhibiting aspect ratios of 1000 or more

Nanoelectromechanical systems (NEMS): refer MEMS

Opto-electronics: an application of electromagnetic photons for electronic device applications; they can be either electrical-to-optical or optical-to-electrical transducers

Photocatalysis: phenomenon of accelerating a chemical reaction rate using a photon beam in the presence of a catalyst

Photoluminescence (PL): a process by which certain substances absorb electromagnetic radiations of specific wavelengths and re-radiate photons of different wavelength

Photonic crystals: periodic dielectric or metallo-dielectric optical nanostructures that are designed to affect the propagation of electromagnetic waves (EM) in the same way as the periodic potential in a semiconductor crystal affects electron motion by defining allowed and forbidden electronic energy bands

Photonics: electronics using light (photons) instead of electrons to manage data

Piezoresistive effect: phenomenon by which electrical resistance of a material varies with externally applied mechanical pressure

Plasma: a state of matter containing a significantly large fraction of ionized matter; plasma properties differ significantly from those of solids, liquids or gases

Physical vapour deposition (PVD): a variety of vacuum deposition technique involving vaporization of atoms from target material to produce a thin film on a substrate

Pyrolysis: Greek word denoting separation (lysis) under fire (pyr); a thermochemical method involving decomposition of organic material at elevated temperatures in the absence of oxygen

Quantum computers: a computational device using quantum mechanical phenomena for operations on input data

Quantum dots: 0D nanostructures in which electron energy states are confined in all three spatial dimensions; their electronic properties are between that of clusters and bulk semiconductors

Qubit: a quantum-computing equivalent to a bit; with an additional dimension of quantum properties of atoms

Resonant tunnelling devices (RTD): 2D quantum devices that consist of a long and narrow semiconductor island, with electron confinement only in two directions

Resonant tunnelling transistors (RTT): see RTD

Scanning near-field optical microscopy (SNOM): illuminates a specimen through an aperture of a size smaller than the wavelength of light used and with the specimen positioned within the near-field regime of the source; by scanning the aperture across the sample through a conventional objective, an image can be formed

Self-assembly: process in which the components interact within themselves to form aligned or organized structures without any external force

Shape memory polymers: smart polymers capable of returning to their original shape after being deformed by external forces, when triggered by an external stimulus such as temperature change

Single electron transistor (SET): devices that are capable of detecting very small variations in the charge of the gate; charge differences of even one electron can cause the on-and-off switching function of SET

Sintering: a powder consolidation technique taking place normally at high temperatures and pressures.

Sol–gel method: a process that involves the generation of a colloidal suspension ('sol'), which is subsequently converted to viscous gel and solid material

Spintronics (spin-based electronics): an emerging technology, which exploits the dual property of electrons, namely charge and spin state; also known as magneto-electronics

Spark plasma sintering (SPS): a sintering technique using pulsed DC current that directly passes through the graphite die, as well as the powder to be consolidated, in case of conductive samples

Superconducting quantum interference device (SQUID): a device capable of measuring extremely weak magnetic fields

Stacking faults: crystallographic defects arising due to wrong stacking sequence of planar arrangement of atoms

Scanning tunnelling microscope (STM): an instrument used for imaging surfaces at the atomic level; it works on the principle of quantum tunnelling

Superplasticity: ability to deform a material well beyond the limits expected from normal tensile tests

Surface plasmon (SP): plasmons that are confined to surfaces and interact strongly with light resulting in a polariton

Targeted drug delivery: administration of a pharmaceutical compound in desired amount to a localized diseased cell/tissue for therapy

Thin film transistors (TFT): an FET made of thin film layers of semiconducting and dielectric materials; used in LCD and digital radiography applications

Thin films: atomically engineered layers with film thickness usually in the range of nanometers to a maximum of a few microns

Tissue engineering: science of structural and functional fundamentals of mammalian tissues and application of biocompatible substitutes to restore, maintain or improve functions

Top-down: involves fragmentation of a microcrystalline material to yield a nanocrystalline material; all solid state synthesis routes of nanostructures fall into this category

Toxicology: study of toxic or poisonous materials and their effect on living systems; the study of biocompatibility and adverse effects of nanomaterials is termed nanotoxicology

Triple junction: a node at the intersection of three crystals or grains

Vapour–liquid–solid method (VLS): a mechanism for the growth of one-dimensional nanostructures, such as nanowires, from chemical vapour deposition; to enhance the efficiency and kinetics for the growth of crystals, a catalytic liquid alloy phase which can rapidly adsorb a vapour to supersaturation levels is used

Whiskers: thin fibrous growth of a dislocation free crystal

X-ray photoelectron spectroscopy (XPS): a quantitative surface chemical analysis technique that measures the elemental composition; the technique involves characterization of photoelectrons produced by irradiating a solid material with x-rays

Zener pinning: pinning of grain boundaries by secondary precipitates, reducing the grain boundary migration kinetics

Zeolite: a material that belongs to the family of hydrous aluminum silicate minerals, containing additions of cations like sodium, potassium, etc., or a corresponding synthetic compound; used chiefly as molecular filters and ion-exchange agents

Index

Fig. 1.3 *(see page 6)* The colours of stained glass panels depend on the particle size of gold and silver nanoparticles introduced into them. Such colloids were made even in the 11th century when medieval artisans mixed gold chloride with glass. (*Source:* http://en.wikipedia.org/wiki/File:Muzeum_Su%C5%82kowskich_-_Zabytkowy_Witra%C5%BC.jpg).

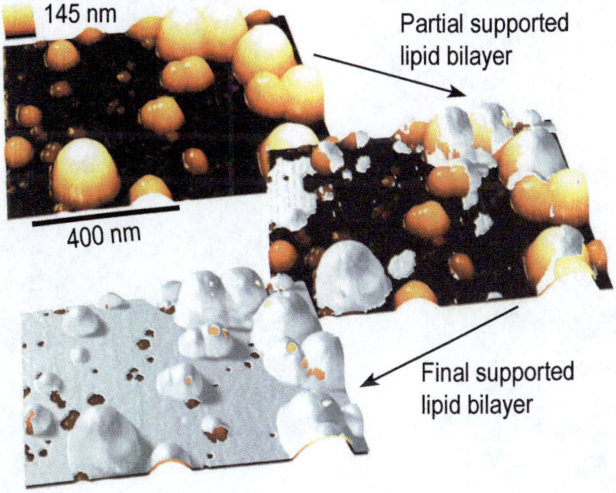

Fig. 1.4 *(see page 6)* Supported lipid bilayer formation on a surface covered with nanoparticles of different sizes. Transmembrane pores (or holes) are formed around particles of approximately 1 to 20 nm diameter. The 3D figures shown here are the adapted representation of the series of tapping mode atomic force microscopy images recorded under liquid buffer medium. (*Source:* http://commons.wikimedia.org/wiki/File:Supported_Lipid_Bilayer_and_Nanoparticles_AFM.png).

Plate 2

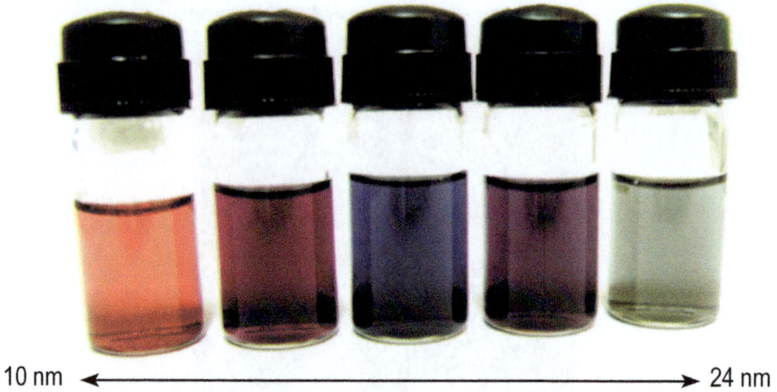

Fig. 1.5 *(see page 7)* Nanoshells designed to absorb various wavelengths of light – including infrared (vial at far right). (*Source:* http://en.wikipedia.org/wiki/File:Nano_Figure_1.png).

Fig. 1.6 *(see page 7)* Famous 4th century Roman cup, which appears green when illuminated from outside and red when illuminated from within; the red colour is due to very small amounts of gold nanopowder (about 40 parts per million). (*Source:* http://commons.wikimedia.org/wiki/File:Lycurgus_cup_british_museum.JPG; http://commons.wikimedia.org/wiki/File:Brit_Mus_13sept10_brooches_etc_046.jpg).

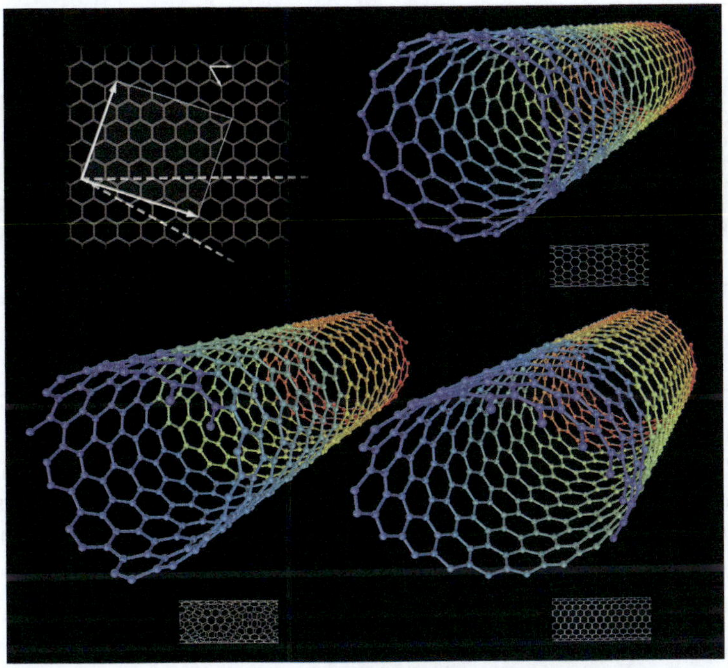

Fig. 1.17 *(see page 16)* Schematic illustration of nanotubes.
(*Source:* http://commons.wikimedia.org/wiki/File:Types_of_Carbon_Nanotubes.png).

(a)

(b)

Fig. 1.21 *(see page 22)* Nanotechnology in nature: self-cleaning lotus leaf. The hydrophobic lotus leaf has inspired nanotechnologists to develop a number of innovative products. (*Source:* (a) http://commons.wikimedia.org/wiki/File:Tropaeolum-majus(Lotus-oben).jpg); (b) http://commons.wikimedia.org/wiki/File:Lotus3.jpg).

Plate 4

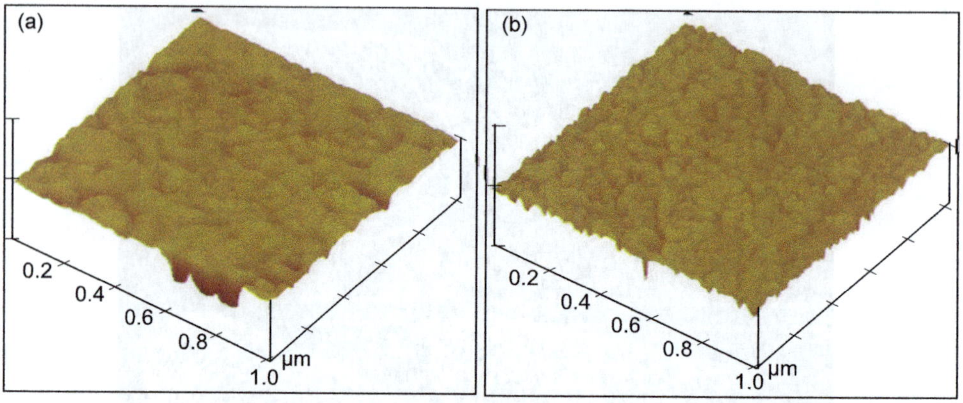

Fig. 2.6 *(see page 42)* Grain growth restriction in Cu–W nanocomposites heated to 500°C in (a) Cu–10%W and (b) Cu–20%W. The as-prepared composite (by high-energy ball milling) shows a grain size of about 20 nm, which is very similar to that observed after heating to 500°C. In comparison, nanocrystalline pure Cu usually shows abnormal grain growth even at 200°C. This clearly brings out the Zener pinning effect in the nanocomposites. (*Source:* BS Murty, IIT Madras)

Fig. 2.19 *(see page 52)* Image of Don Quixote. Similar figures have been made using nanocrystalline VO_2. The image is invisible below 341K. The visibility in the picture is due to a drastic change in optical property associated with phase transition in nanocrystalline VO_2. This is reported to be the world's fastest optical shutter known till date. (*Source:* http://upload.wikimedia.org/wikipedia/commons/3/32/Honor%C3%A9_Daumier_017.jpg; http://www.vanderbilt.edu/exploration/stories/vo2shutter.html (Credit: René Lopez, Vanderbilt University))

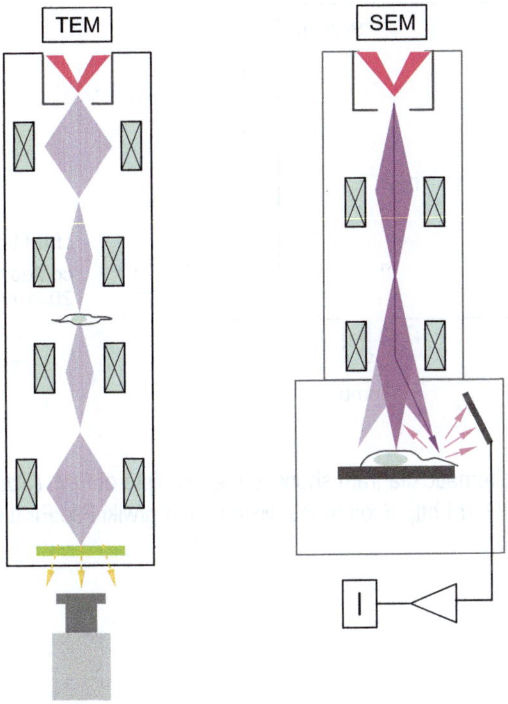

Fig. 5.4 *(see page 154)* Ray diagrams for SEM and TEM.
(*Source:* http://commons.wikimedia.org/wiki/File:SimpleSEMandTEM.jpg)

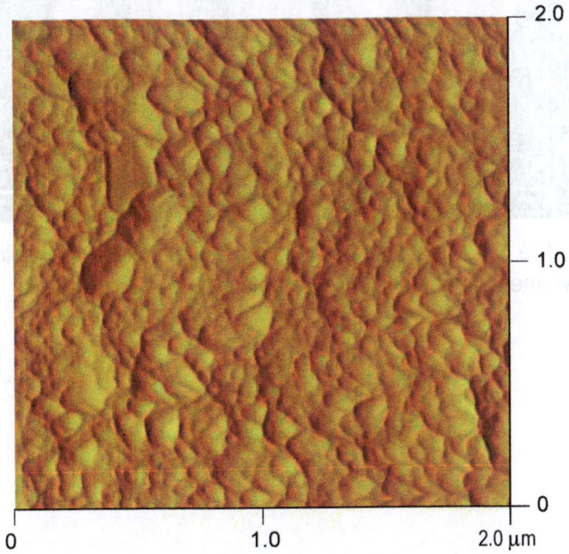

Fig. 5.10 *(see page 165)* AFM image of a Cu–Ta nanocomposite. (*Source:* BS Murty, IIT Madras).

Plate 6

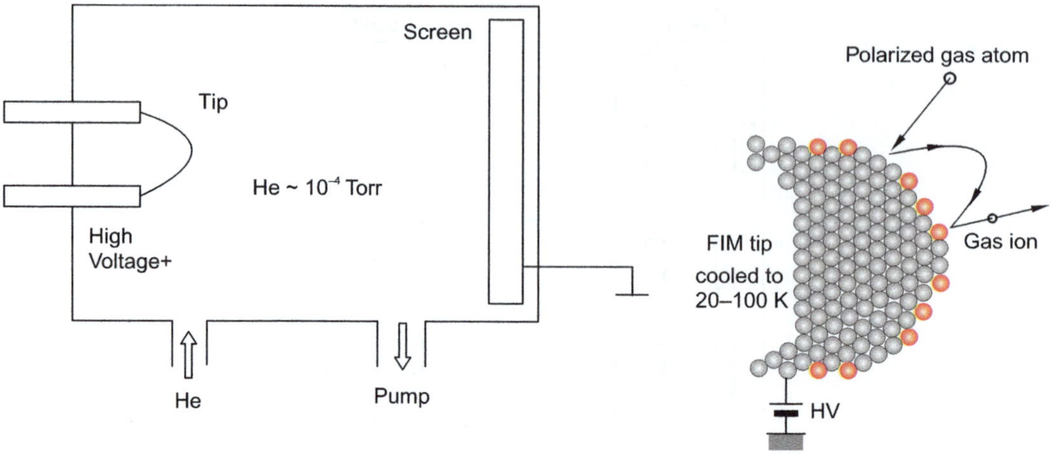

Fig. 5.12 *(see page 169)* Schematic diagram showing the principle of FIM. (*Source:* (a) http://commons.wikimedia.org/wiki/File:FIM.JPG; (b) http://commons.wikimedia.org/wiki/File:FIMtip.JPG)

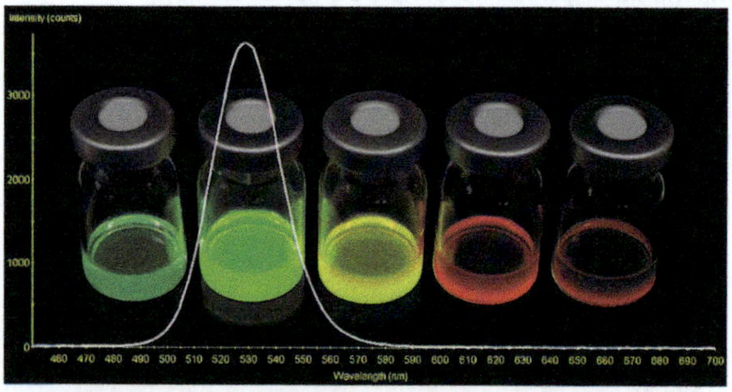

Fig. 6.2 *(see page 178)* CdSe NCQDs that fluoresce into different colours depending on their size. (*Source:* http://commons.wikimedia.org/wiki/File:CdSeqdots.jpg).

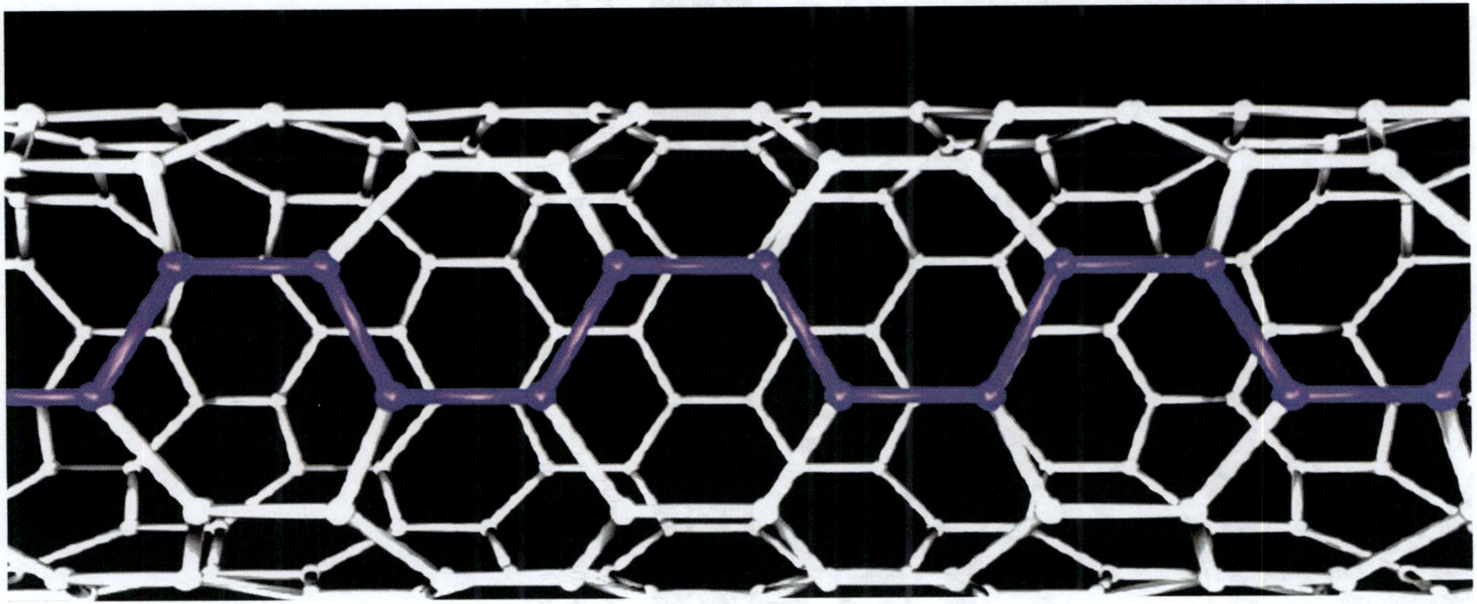

Fig. 6.7 *(see page 183)* Ball–stick model of a nanotube; the balls represent the carbon atoms and the sticks their bonds. (*Source:* http://commons.wikimedia.org/wiki/File:Semiconducting_nanotube.png)

Plate 8

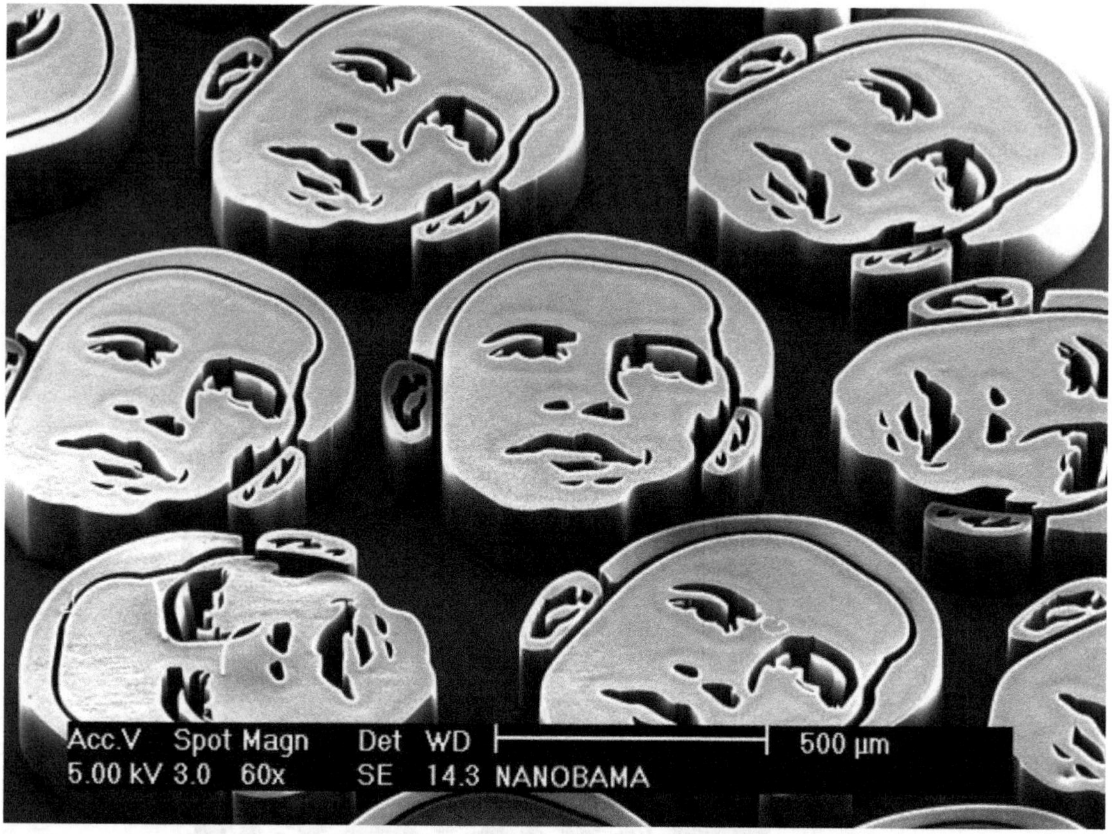

Image of Barak Obama produced using nanoparticles. Each face is made of approximately 150 million tiny carbon nanotubes; that's about how many Americans voted in the 2008 presidential election.
(With kind permission from John Hart, Sameh Tawfick, Michael De Volder from the University of Michigan, USA*)*

9783642280290